5th International Conference on Radiotherapy Gel Dosimetry 2008

Journal of Physics: Conference Series Volume 164

Hersonissos, Crete, Greece
29 September – 3 October 2008

Editors:

Thomas G. Maris Evangelos Pappas

ISBN: 978-1-61738-370-0

Printed from e-media with permission by:

Curran Associates, Inc.
57 Morehouse Lane
Red Hook, NY 12571

Printed by Curran Associates, Inc. (2010)

For permission requests, please contact the Institute of Physics
at the address below.

Institute of Physics
Dirac House, Temple Back
Bristol BS1 6BE UK

Phone: 44 1 17 929 7481
Fax: 44 1 17 920 0979

techtracking@iop.org

Additional copies of this publication are available from:

Curran Associates, Inc.
57 Morehouse Lane
Red Hook, NY 12571 USA
Phone: 845-758-0400
Fax: 845-758-2634
Email: curran@proceedings.com
Web: www.proceedings.com

TABLE OF CONTENTS

DOSIMETRY HISTORICAL REVIEW

CHEMISTRY AND FUNDAMENTAL PROPERTIES OF POLYMER GEL DOSIMETERS

GEL DOSIMETRY WITH OPTICAL COMPUTED TOMOGRAPHY

GEL DOSIMETRY WITH MAGNETIC RESONANCE IMAGING

GEL DOSIMETRY WITH OTHER THAN OPTICAL CT AND MR SCAN TECHNIQUES

OTHER 3D DOSIMETERS

GEL DOSIMETRY APPLICATIONS (1)

GEL DOSIMETRY APPLICATIONS (2)

The International Conference on Radiotherapy Gel Dosimetry (DOSGEL) is held every two years. Its purpose is to bring together basic science and clinical researchers, medical physicists and clinicians from around the world to discuss the state-of-the-art of the gel dosimetry technique and to set the directions and trends for its future improvements.

Gel dosimetry can be broadly defined as using a gel that can react to the absorption of ionizing radiation, and that can retain this information which can subsequently be retrieved by an external imaging modality. Examples of radiation-sensitive gels include, but are not limited to, polymer gel dosimeters, Fricke gel dosimeters and others. Imaging modalities that are of general use in this field are (in alphabetical order) magnetic resonance imaging (MRI), optical light computed tomography and x-ray computed tomography.

This volume comprises the proceedings of the 5th International Conference on Radiotherapy Gel Dosimetry (DOSGEL 2008). The conference, organised by the University of Crete, Medical Physics Department, took place in Hersonissos, Crete, Greece from 29 September to 3 October 2008. The meeting aimed to continue the series of biannual DOSGEL conferences and focused on the promotion of gel dosimetry techniques by setting the trends for their future improvements.

The main scientific session topics of DOSGEL 2008 were the following:

- Chemistry and fundamental properties of polymer gel dosimeters
- Gel dosimetry with Optical Computed Tomography
- Gel dosimetry with Magnetic Resonance Imaging
- Gel dosimetry with other than Optical CT and MR scan Techniques
- Other 3D dosimeters
- Gel dosimetry applications

Local Organizing Committee

Thomas G Maris (University of Crete, Greece, Chairman DOSGEL 2008)
John Damilakis (University of Crete, Greece)
Evangelos Pappas (University of Crete, Greece)
Antonios Papadakis (University of Crete, Greece)
Fotini Zacharopoulou (University of Crete, Greece)
John Stratakis (University of Crete, Greece)
Pantelis Karaiskos (University of Athens, Greece)
Panos Papagiannis (University of Athens, Greece)

Scientific Committee

President: Yves De Deene (Ghent University, Belgium)
Sven Back (Lund University, Sweden)
Clive Baldock (University of Sydney, Australia)
David Bonnett (Kent Oncology Center, UK)

Simon Doran (University of Surrey, UK)
Cheryl Duzenli (University of British Columbia, Canada)
Geoffrey Ibbott (Colorado State University, USA)
Andrew Jirasek (University of Victoria, Canada)
Kevin Jordan (University of Western Ontario, Canada)
Martin Lepage (Universite de Sherbrooke, Canada)
Mark Oldham (Duke University, USA)
L John Schreiner (Kingston Regional Cancer Centre, Canada)

Acknowledgements
The local organising committee wishes to express its gratitude to all participants for their activities at DOSGEL 2008 and for creating such a friendly and inspiring environment. Special thanks are due to all the speakers, for preparing and presenting their talks, and for many valuable discussions. We also give thanks to all members of the scientific committee who, acting as referees, improved significantly the scientific quality of this proceedings volume. We would also like to thank all chairmen for their efficient leading of sessions.

On Behalf of the local organizing committee of DOSGEL 2008

Thomas G Maris and Evangelos Pappas
Editors

Journal of Physics: Conference Series **164** (2009) 012001 doi:10.1088/1742-6596/164/1/012001

Where does gel dosimetry fit in the clinic?

L.J. Schreiner

Departments of Oncology and Physics, Queen's University, and Cancer of Centre Southeastern Ontario at KGH, Kingston, Canada, K7L 5P9

john.schreiner@krcc.on.ca

Abstract. For over a decade our community has considered gel dosimetry a promising technique for the clinic, since it provides a unique methodology for three dimensionally dose measurement of the complex conformal dose distributions achieved by modern techniques such as Intensity Modulated Radiation Therapy. Even with improved gel dosimeters (such as normoxic polymer gel systems) and more accessible imaging for dose readout, this potential has not yet been fully realised. This may be because alternative tools have been able to more easily provide the technical and dosimetry quality assurance performed at specific steps in radiation treatment. The development of Adaptive Radiation Therapy is introducing complex processes for radiation treatment that depend on intricate links between various stages from imaging to treatment planning to radiation delivery, and back through loops depending to the level of adaptation. These links often involve human interpretation and intervention. Therefore, there is a greater need for comprehensive quality assurance of the whole radiation therapy process when adaptive techniques are first being developed, in regular quality assurance of the performance of these techniques within a radiation oncology program, and in independent outside credentialing of cancer centres by independent agents. I will attempt to make the case in this introductory paper to Dosgel08 that this is the role uniquely served by gel dosimetry.

1. Introduction

Radiation therapy is a dominant cancer therapy with an appropriate level use estimated in excess of 50% in developed countries [1,2]. Radiotherapy is a localized treatment intended to deliver a sufficient and uniform dose to the target, to achieve tumour control, while minimizing complications by limiting the dose to normal tissue. Over the years, external radiation therapy has progressed considerably towards this goal by the development of three dimensional (3D) conformal radiation therapy techniques [3,4] such as achieved by Intensity Modulated Radiation Therapy (IMRT). These are becoming standards of practice benefiting a significant proportion of patients [4-6]. Conformal therapy by brachytherapy has also advanced through improvements in delivery systems and in imaging for planning and for source localization.

The dose distributions resulting from 3D conformal therapy and IMRT are designed to fit tightly about the specific target volumes and, therefore, the possibility of missing the target increases, because of patient setup errors, organ motion, or even small fluctuations in treatment delivery. Therefore, the verification of the actual dose delivery is vital to the effective delivery of conformal radiation therapy [4,7-9]. The conformal dose delivery achieved in modern IMRT delivery techniques results in complex dose distributions characterized by non-uniform doses with large dose gradients in small spatial displacement. Furthermore, the doses are delivered dynamically, so that specific points in the irradiated volumes may receive their final dose only over a total treatment time. Thus the dose measurement required for the verification of IMRT dose delivery becomes very labour intensive and often impractical with traditional dosimetry (e.g., using ion chamber, TLD, or film techniques) [10-12]. There is, therefore, great interest in developing new verification techniques that provide improved dosimetry techniques using flat panel devices, 2D and 3D arrays of detectors, scintillation devices and other means [9,13-15].

The limitations of conventional dosimeters spurred the development of gel dosimetry [7,16-18], the topic of this workshop. Gel dosimetry has shown considerable promise for high resolution, tissue equivalent 3D dosimetry for nearly two decades (see Figure 1). However, it has still not come into common clinical use. Initially this was the result of two problems: 1) the original polymer gel dosimeters (which had greater temporal stability) were difficult to prepare in the hospital setting as they contained toxic constituents and, until recently, required anoxic preparation and 2) ready access to imaging techniques for clinical reading of the gels was often very restricted. Recent developments suggest that these impediments are slowly being removed. The development of normoxic and less toxic polymer gel dosimeters, and the development of optical imaging techniques for readout in Fricke gel dosimetry, have made their use in the clinical setting more tenable. And, while MRI has historically been the standard imaging modality [17,19-21] for quantitative 3-D dose distributions, other physical properties of irradiated gel, such as opacity and density, also vary with dose as a result of the radiation-induced formation and precipitation of polymer. This provides the potential for dose imaging using optical techniques [22-25], or x-ray computed tomography (CT) [26-28]. These alternative imaging schemes are attractive since access to MRI, the gold standard imaging, is often very limited given the long waiting lists for patients requiring diagnostic scans [27,28]. Optical CT imaging (OptCT, to distinguish the technique from OCT, optical coherence tomography) is especially promising because of the ability to construct accurate low-cost benchtop scanners [22-25,29,30].

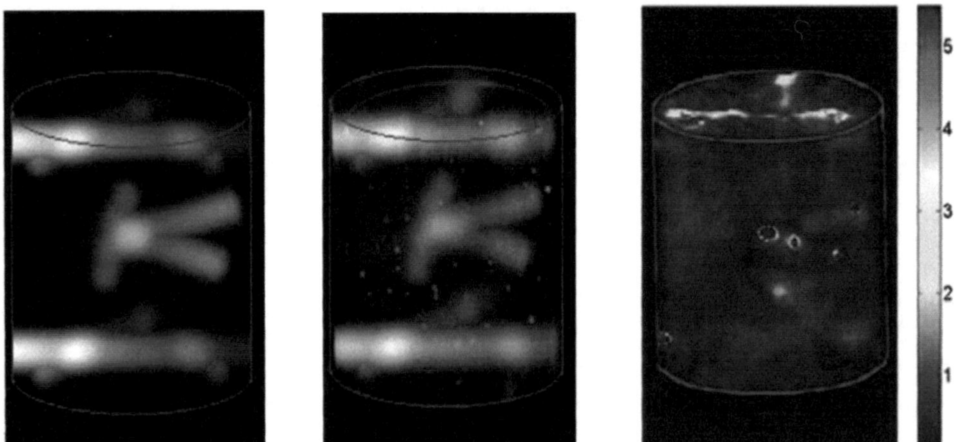

Figure 1 - An illustration of the strength of gel dosimetry from an example with pencil beam irradiations on a bench top cobalt tomotherapy device[31]. The left hand image shows a maximum intensity projection (MIP) dose map of the 3-D dose plan generated by our in house treatment planning system. The middle figure is the MIP [32,33] image for the same dose delivery as determined by optical CT measurements on a VISTA cone beam CT system. The right hand image is a three dimensional 5% 5mm gamma evaluation of the Low gamma function [34,35]. In this particular case we have used the gel dosimeter to assess/commission our treatment planning system. From the location of the failures we have been able to identify limitations with the planning system algorithm.

Even with these improvements, gel dosimetry has not achieved the clinical prominence indicated by its promise in terms of resolution, tissue equivalence, etc. This may be because in many instances its practical use is more laborious than really required, especially in situations well handled by other techniques. For example, while gel dosimetry does provide a strong modality for the quality assurance of IMRT delivery, other techniques such as the use of film[36-38], dosimeter arrays[39-42] and portal imaging[43-45] provide sufficient assurance of dose delivery for the day to day clinical practice. Then perhaps the strength of gel dosimetry is not the evaluation of 'familiar' IMRT delivery (i.e., the delivery that has been well tested and established in a clinic).

2. Image Guided and Adaptive Radiation Therapy

The requirement for accurate targeting in conformal therapy was recently indicated by de Crevoisier et al. in a retrospective analysis of treatment plans for radically treated prostate cancer patients [46]. The authors revealed that the extent of rectal distension in their planning scans was a predictor for biochemical control. The observed result of target miss (since on treatment without rectal filling the prostate would have shifted) was that the biochemical control rates in the distended-rectum population were 60% (all stages) as opposed to 83% in the non-distended population.

This work serves to underscore the previous recognition of the importance to develop treatment setup and image guidance techniques [47,48]. On-line imaging (particularly cone beam CT) is prominent in this development of Image Guided Radiation Therapy (IGRT) [49-56]. IGRT is inherent in helical tomotherapy units [48,57] and has been augmented on conventional linacs by advances in x-ray detector technology and computing capacity [55,56,58-61]. IGRT is becoming a standard of care in radiation therapy [62] with over three quarters of the new radiation therapy units being installed with cone-beam CT systems for guidance. It should be noted that some do debate the value of IGRT with current technology[58].

One observation from the increased use of IGRT has been to note the potential for significant re-modeling of the tumour and surrounding tissue during the time course of the multiple fractions of radiotherapy[63,64]. This change of anatomy which might be seen over the weeks of IGRT has lead to the development of off-line approaches wherein the treatment plan is re-designed (daily or weekly) to accommodate volume changes ensuring consistent target coverage or avoidance of normal structures. Such Adaptive Radiation Therapy (ART) may incorporate kVCT (on an IGRT linac) or MVCT (on a tomotherapy unit) images and treatment monitoring and potentially for treatment planning (although the patient may be directed back to the CT simulator for replanning, see Figure 2). This process requires considerable data processing for image registration and analysis[62,65-67]. More sophisticated approaches envisage adaptation over each fraction [68]. The proposals for inter-fraction ART are not limited to corrections for tissue geometry. The potential for on-line dosimetry (perhaps via exit beam dosimetry with portal images or tomotherapy CT detectors) during each treatment raises the feasibility of dose delivery monitoring with subsequent correction as fractions continue through the treatment course.[69-73]. This implies that each fraction may delivery slightly different doses via slightly different MLC sequencing. Other researchers have proposed using biological and functional imaging to assess tumour response (perhaps looking at changes in hypoxic tissue as treatment progresses). These sophisticated plans require a change of subsequent dose delivery and the efficacy of the treatment will require that the dose is delivered correctly after all of these interventions. Therefore, the dosimetry problem initially introduced with the development of conformal therapy has become more complex still. Modern treatment is an increasingly complex process involving data transfer between sophisticated advanced equipment, often with human evaluation at critical steps to evaluate and choose between different courses of action at critical steps.

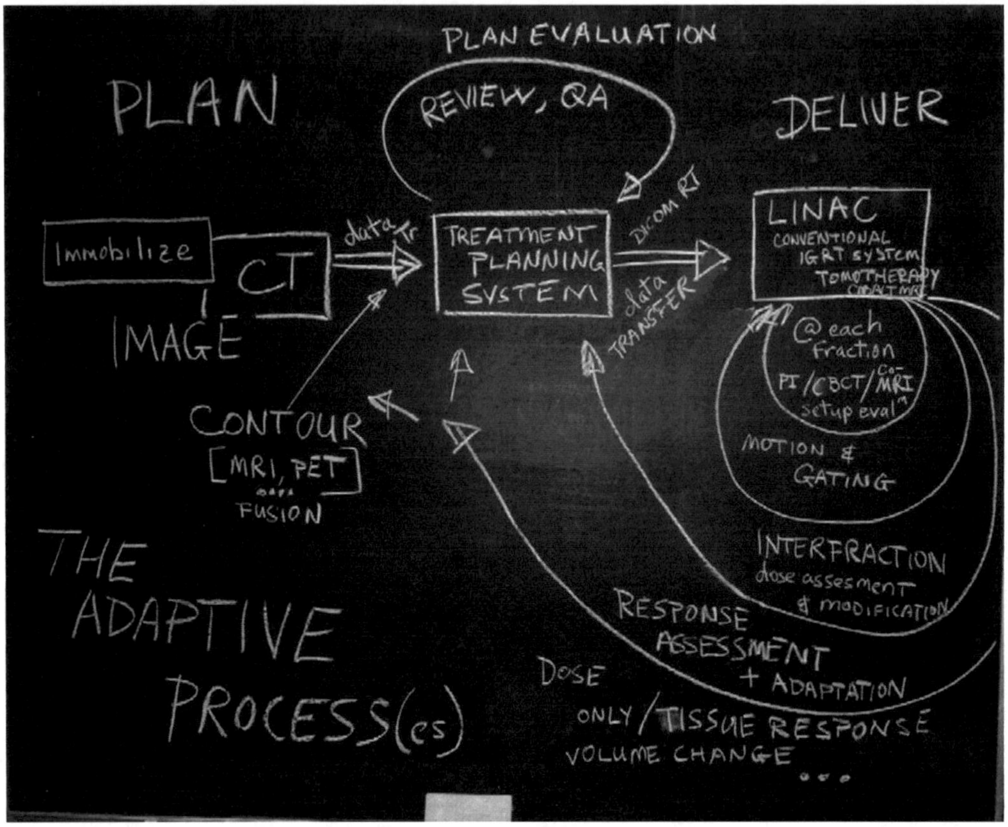

Figure 2 - A diagram showing some of the potential processes that are in current practice or have been envisaged for various strategies of adaptive radiation therapy. In general there is the common thread going from initial imaging for planning purposes (which may involve multiple modalities and image fusion), contouring by a physician or planning therapist, treatment planning by a planning therapist or dosimetrist with plan evaluation and review by physicists and radiation oncologists, data transfer to the treatment unit which has some kind of image guidance and perhaps exit dosimetry available and then treatment delivery. The adaptive process may be as simple as on-board imaging immediately prior to treatment with megavoltage portal imaging or kV imaging (perhaps cone beam CT) for setup validation with correction if required before treatment. As schemes get more sophisticated they may involve correction for patient motion perhaps by gating the radiation beam so that it is irradiating the patient when the target is in the beam port, or by dynamically collimating the radiation so that it follows the moving target of the beam to accommodate organ motion or even respiratory targeting in which the multi-leaf collimators are tracking the respiratory motion. This second approach may be more manageable in units in which the target is observed directly rather that the motion followed through surrogate measurements. This is one motivation for the proposed cobalt MRI units [70,71] which may enable imaging during the radiation delivery. In various schemes of inter-fraction adaptive therapy subsequent radiation delivery may be adjusted to ensure the dose delivery over a complete course is correct (based on comparison of daily doses determined by exit dose measurement to intended) of the account for biological response to treatment (either from changes in tissue/organ geometry in the treatment field to assessment of response by biological imaging). In each of these schemas, the processes are complicated and require sophisticated technology along with human interpretation and intervention. The quality assurance of these processes is not easily achieved by traditional QA tools and tests which usually focus at limited select points in the process.

3. Modern Radiation Therapy QA: From Dosimetry to Process validation

How well then is the radiation oncology community assuring that modern radiation therapy is well delivered? As noted above, there are a large number of techniques that have been established for dosimetric verification at specific stages of IMRT (for example, the assessment of individual port delivery through a given sequence of dynamic MLC at a given gantry angle). Similarly, various approaches have been proposed for the quality assurance of other technical components involved at different stages of IGRT [74-80]. Other workers have reviewed the particular experience of their QA programs perhaps to evaluate the impact of new technology, or to establish benchmark tolerances for deviations in particular tests[81-88].

An indication of the success of our efforts has been reported recently by Geoff Ibbott and colleagues at the Radiological Physics Centre. While credentialing institutions participating in IMRT clinical trials under the auspices of the National Cancer Institute, the RPC provided them with anthropomorphic phantoms (of the head and pelvis) containing various dosimeters and regions of interest for irradiation. The centres were given an IMRT objective and asked to irradiate the phantom using their usual IMRT planning and delivery. The results were sobering; about one third of the centres failed to achieve the goal of irradiating correctly within 7% dose or 4mm distance in the high dose region near the organ at risk.

The RPC experience, the results of analyses by various individuals in their clinics, and the deliberations of working groups are leading some to conclude that radiation therapy quality assurance programs need to be rethought. For example, people are suggesting that risk analysis be undertaken to identify critical points requiring greater vigilance. Also there is an appreciation that new approaches are required to look at the complete chain inherent to modern delivery accounting for the complex interactions and connections of the various components [89-92]. With this view it may be that clinical QA will become more of an integrated activity and I would advocate that gel dosimetry may have a unique role to play in this future.

4. Clinical role(s) for gel dosimetry

It is clear from the discussion above and from the illustration in Figure 2, that IGRT is a complex process and that the treatment of a particular patient with adaptive approaches requires considerable effort. I would suggest that one way to assess if the process is working well is to use gel dosimetry to mimic a patient throughout all the process including the various adaptive steps. Similar schemas that have been proposed over the years illustrating the strength of gel dosimetry for quality assurance of IMRT [12,93-96] and of precision stereotactic delivery [97]. Recently a couple of groups [98-100] have shown that gel and radiochromic solid polymer dosimeters operate well in the RPC phantom. Preliminary investigation has also shown that gel dosimeters can be used in conjunction with dynamic phantoms[101]. In all these examples, the gel dosimeter mimics the target and surrounding volume sufficiently well that contouring and imaging are possible and the complete planning and delivery validated.

The same approach is readily adapted to the evaluation of ART in a given clinic (see Figure 3). With suitably stable dosimeters, the tests could be run over multiple fractions to validate inter-fraction adaptation. While the dosimeter and phantom preparation should be done by the physics group (or perhaps a credentialing service), the rest of treatment planning and delivery should be performed by the individuals taking on their various roles as in patient treatment (e.g., CT imaging by planning therapists, contouring by oncologists, computerised treatment planning by the planning RTs and dosimetrists, plan evaluation by physicists and oncologists, treatment setup, onboard imaging and delivery by the treatment therapists, etc). Then the tool would be handed over to the physicists at the end of the process for the final confirmation that the intended dose delivery was achieved. The intent is to perform the quality assurance of the complete adaptive flow for the treatment of the patient in the clinic. There will be further tests required if the gel dosimetry fails. However, good performance at regular intervals would provide a measure of confidence in the operation of the comprehensive program. While this application of gel dosimetry is not much different than that envisaged since the

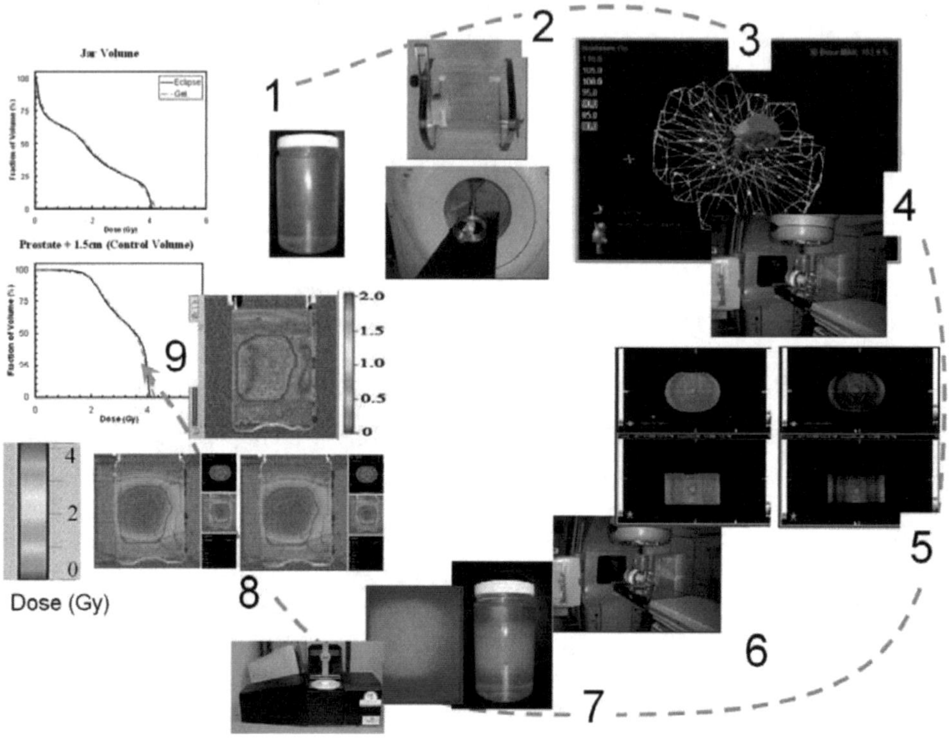

Figure 3 – An example of the implementation of a gel dosimetry quality assurance assessment of an Adaptive Radiotherapy Procedure. In this particular case the work is to QA the simple case of cone beam kVCT verification and correction of patient setup prior to treatment. In the process QA advocated in this paper various steps could include: 1) preparation of the sample and insertion of sample into an appropriate phantom (preferably anthropomorphic) by the physics team, 2) imaging and contouring of images by the CT therapists and radiation oncologist, 3) treatment planning by the dosimetrist /planning therapist then transfer of the beam data and phantom images to the treatment unit, 4) set up verification in this case with the cone beam CT available on the linear accelerator, 5) adjustment of patient setup (in this case through a automated table shift determined by the Varian registration software as per clinical practice, 6) dose delivery as per the treatment plan, (steps 4-6 would be performed by treatment therapists) 7) subsequent removal of the dosimeter and dose imaging by the physics team and 8) data registration and 9) test evaluation by the physicists. In this image the image on the left at 8) is the treatment plan and the image on the right is the measurement. The colour was above is the gamma evaluation, our software enable comparison of the dose volume histograms for any contour specified.

early days of gel dosimetry, the change is that the verification is not limited to testing individual steps in the treatment process, but rather in the validation of the whole process.

5. Conclusions

The last few years have continued to push gel dosimetry into the clinic. New dosimeter formulations and more readily available imaging have made its use more attractive. Along with the internal ART and IGRT quality assurance that I am advocating, many of the uses suggested in the past are still valid. Gel dosimetry is well suited for the commissioning of treatment delivery and treatment planning in difficult situations that may require high resolution data, and the commissioning of new treatment techniques using an approach similar to Figure 3 but executed by an implementation team.

We will hear about these uses over the next four days of this workshop, and I am looking forward to learning how we can better use the gels that we have been so passionate about for these many years. The take home message I want leave you with at the onset is that we have too often advocated the use of gel dosimeters when other tools were much more suitable. This may have hurt the adoption of the technique. I believe gel dosimetry will finally come to fruition when we advocate its use where other tools cannot serve.

6. Acknowledgements

Financial support has been provided by the Canada Institutes for Heath Research (CIHR). Thanks to Tim Olding and Oliver Holmes for their valuable assistance in preparing this introduction and their outstanding work to illustrate the potential of gel dosimetry in our clinical setting.

7. References

[1] Barbera L, Zhang-Salomons J, Huang J, Tyldesley S and Mackillop W 2003 Defining the need for radiotherapy for lung cancer in the general population: a criterion-based, benchmarking approach. *Med Care* **41** 1074-85

[2] Kerba M, Miao Q, Zhang-Salomons J and Mackillop W 2007 Defining the need for breast cancer radiotherapy in the general population: a criterion-based benchmarking approach. *Clin Oncol (R Coll Radiol)* **19** 481-9

[3] Webb S 1993 The Physics of Three-Dimensional Radiation Therapy. Mould R F, Orton C G, Spaan J A E and Webster J G, editors. (Bristol, UK, IOP Publishing Ltd)

[4] IMRT Working Group 2001 Intensity Modulated Radiotherapy: Current Status and Issues of Interest. *Int J Radiat Oncol Biol Phys* **51** 880-914

[5] Donaldson S S and Boyer A L 2000 New methods for precision radiation therapy exceed biological and clinical knowledge and institutional resources needed for implementation. *Med Phys* **27** 2477-9

[6] Keall P J and Williamson J F 2003 Clinical evidence that more precisely defined dose distributions will improve cancer survival and decrease morbidity. *Med Phys* **30** 1281-2

[7] Kron T. Gel Dosimetry: Needs, Expectations, and Reality. In: Schreiner LJ, Audet C, editors. Proceedings of the 1st International Workshop on Radiation Therapy Gel dosimetry. Edmonton: Canadian Organization of Medical Physicists, 1999.

[8] De Wagter C 2004 The ideal dosimeter for intensity modulated radiation therapy (IMRT): What is required? *Journal of Physics: Conference Series* **3** 4-8

[9] Petric M P, Robar J L and Clark B G 2006 Development and characterization of a tissue equivalent plastic scintillator based dosimetry system. *Med Phys* **33** 96-105

[10] Low D A, Mutic S, Dempsey J F, Gerber R L, Bosch W R, Perez C A et al. 1998 Quantitative dosimetric verification of an IMRT planning and delivery system. *Radiother Oncol* **49** 305-16

[11] Paliwal B R, Ritter M A, McNutt T R, Mackie T R, Thomadsen B R, Purdy J A et al. 1998 A solid water pelvic and prostate phantom for imaging, volume rendering, treatment planning, and dosimetry for an RTOG multi- institutional, 3-D dose escalation study. Radiation Therapy Oncology Group. *Int J Radiat Oncol Biol Phys* **42** 205-11

[12] Sandilos P, Angelopoulos A, Baras P, Dardoufas K, Karaiskos P, Kipouros P et al. 2004 Dose verification in clinical IMRT prostate incidents. *International Journal of Radiation Oncology, Biology, Physics 59(5):1540-7,*

[13] Archambault L, Arsenault J, Gingras L, Beddar A S, Roy R and Beaulieu L 2005 Plastic scintillation dosimetry: optimal selection of scintillating fibers and scintillators. *Med Phys* **32** 2271-8

[14] Moran J M, Roberts D A, Nurushev T S, Antonuk L E, El Mohri Y and Fraass B A 2005 An Active Matrix Flat Panel Dosimeter (AMFPD) for in-phantom dosimetric measurements. *Med Phys* **32** 466-72
[15] Archambault L, Beddar A S, Gingras L, Roy R and Beaulieu L 2006 Measurement accuracy and cerenkov removal for high performance, high spatial resolution scintillation dosimetry. *Med Phys* **33** 128-35
[16] Gore J C, Kang Y S and Schulz R J 1984 Measurement of radiation dose distributions by nuclear magnetic resonance (NMR) imaging. *Phys Med Biol* **29** 1189-97
[17] Maryanski M J, Gore J C, Kennan R P and Schulz R J 1993 NMR relaxation enhancement in gels polymerized and cross-linked by ionizing radiation: a new approach to 3D dosimetry by MRI. *Magn Reson Imaging* **11** 253-8
[18] Schreiner LJ. Gel Dosimetry: Motivation and Historical Foundation. In: Schreiner LJ, Audet C, editors. Proceedings of the 1st International Workshop on Radiation Therapy Gel dosimetry. Edmonton: Canadian Organization of Medical Physicists, 1999: 1-10.
[19] Maryanski M J, Ibbott G S, Eastman P, Schulz R J and Gore J C 1996 Radiation therapy dosimetry using magnetic resonance imaging of polymer gels. *Med Phys* **23** 699-705
[20] Baldock C, Lepage M, Back S A, Murry P J, Jayasekera P M, Porter D et al. 2001 Dose resolution in radiotherapy polymer gel dosimetry: effect of echo spacing in MRI pulse sequence. *Phys Med Biol* **46** 449-60
[21] De Deene Y, De Wagter C, Van Duyse B, Derycke S, De Neve W and Achten E 1998 Three-dimensional dosimetry using polymer gel and magnetic resonance imaging applied to the verification of conformal radiation therapy in head-&-neck cancer. *Radiother Oncol* **48** 283-91
[22] Doran S J, Koerkamp K K, Bero M A, Jenneson P, Morton E J and Gilboy W B 2001 A CCD-based optical CT scanner for high-resolution 3D imaging of radiation dose distributions: equipment specifications, optical simulations preliminary results. *Phys Med Biol* **46** 3191-213
[23] Oldham M, Siewerdsen J H, Shetty A and Jaffray D A 2001 High resolution gel-dosimetry by optical-CT and MR scanning. *Med Phys* **28** 1436-45
[24] Oldham M, Siewerdsen J H, Kumar S, Wong J and Jaffray D A 2003 Optical-CT gel-dosimetry I: basic investigations. *Med Phys* **30** 623-34
[25] Oldham M and Kim L 2004 Optical-CT gel-dosimetry. II: Optical artifacts and geometrical distortion. *Med Phys* **31** 1093-104
[26] Audet C, Hilts M, Jirasek A and Duzenli C 2002 CT gel dosimetry technique: comparison of a planned and measured 3D stereotactic dose volume. *Journal of Applied Clinical Medical Physics* **3** 110-8
[27] Hilts M, Audet C, Duzenli C and Jirasek A 2000 Polymer gel dosimetry using x-ray computed tomography: a feasibility study. *Phys Med Biol* **45** 2559-71
[28] Hilts M, Jirasek A and Duzenli C 2005 Technical considerations for implementation of x-ray CT polymer gel dosimetry. *Physics in Medicine & Biology* **50** 1727-45
[29] Kelly B G, Jordan K J and Battista J J 1998 Optical CT reconstruction of 3D dose distributions using the ferrous- benzoic-xylenol (FBX) gel dosimeter. *Med Phys* **25** 1741-50
[30] Maryanski M J, Zastavker Y Z and Gore J C 1996 Radiation dose distributions in three dimensions from tomographic optical density scanning of polymer gels: II. Optical properties of the BANG polymer gel. *Phys Med Biol* **41** 2705-17
[31] Schreiner LJ, Kerr AT, Salomons GJ, Dyck C, Hajdok G. The Potential For Image Guided Radiation Therapy With Cobalt-60 Tomotherapy. Proc. 6th Annual International Conference on Medical Image Computing and Computer Assisted Intervention. Heidelberg, Germany: Springer, 2003: 449-456.
[32] Wallis J W and Miller T R 1991 Three-dimensional display in nuclear medicine and radiology. *J Nucl Med* **32** 534-46
[33] Wallis J W, Miller T R, Lerner C A and Kleerup E C 1989 Three-dimensional display in nuclear medicine. *IEEE Trans Med Imaging* **8** 297-30

[34] Low D A, Harms W B, Mutic S and Purdy J A 1998 A technique for the quantitative evaluation of dose distributions. *Med Phys* **25** 656-61
[35] Low D A and Dempsey J F 2003 Evaluation of the gamma dose distribution comparison method. *Med Phys* **30** 2455-64
[36] Kapulsky A, Gejerman G and Hanley J 2004 A clinical application of an automated phantom-film QA procedure for validation of IMRT treatment planning and delivery. *Med Dosim* **29** 279-84
[37] Gillis S and De Wagter C 2005 Practical and dosimetric implications of a new type of packaging for radiographic film. *Phys Med Biol* **50** N63-N72
[38] Kron T, Yartsev S and Mackie T R 2005 Verification dosimetry during treatment for helical tomotherapy using radiographic film. *Australas Phys Eng Sci Med* **28** 232-7
[39] Jursinic P A and Nelms B E 2003 A 2-D diode array and analysis software for verification of intensity modulated radiation therapy delivery. *Med Phys* **30** 870-9
[40] Langen K M, Meeks S L, Poole D O, Wagner T H, Willoughby T R, Zeidan O A et al. 2005 Evaluation of a diode array for QA measurements on a helical tomotherapy unit. *Med Phys* **32** 3424-30
[41] Letourneau D, Gulam M, Yan D, Oldham M and Wong J W 2004 Evaluation of a 2D diode array for IMRT quality assurance. *Radiother Oncol* **70** 199-206
[42] Poppe B, Blechschmidt A, Djouguela A, Kollhoff R, Rubach A, Willborn K C et al. 2006 Two-dimensional ionization chamber arrays for IMRT plan verification. *Med Phys* **33** 1005-15
[43] Gagel B, Schramm O, Harms W, Mulhern A, Wenz F, Van Kampen M et al. 2002 The electronic portal imaging system Siemens Beamview Plus versus the conventional verification films CEA-TVS and DuPont COL-7. A critical appraisal of visual image quality. *Strahlenther Onkol* **178** 446-52
[44] Norrlinger B, Islam M K, Heaton R, Fan E and Ramaseshan R 2003 Relative dosimetry using an amorphous silicon electronic portal imaging device for the quality assurance of dynamic intensity modulated radiation therapy. *Int J Radiat Oncol Biol Phys* **57** S423
[45] Grein E E, Lee R and Luchka K 2002 An investigation of a new amorphous silicon electronic portal imaging device for transit dosimetry. *Med Phys* **29** 2262-8
[46] de Crevoisier R, Tucker S L, Dong L, Mohan R, Cheung R, Cox J D et al. 2005 Increased risk of biochemical and local failure in patients with distended rectum on the planning CT for prostate cancer radiotherapy. *Int J Radiat Oncol Biol Phys* **62** 965-73
[47] Low D A, Grigsby P W, Dempsey J F, Mutic S, Williamson J F, Markman J et al. 2002 Applicator-guided intensity-modulated radiation therapy. *Int J Radiat Oncol Biol Phys* **52** 1400-6
[48] Mackie T R, Kapatoes J, Ruchala K, Lu W, Wu C, Olivera G et al. 2003 Image guidance for precise conformal radiotherapy. *Int J Radiat Oncol Biol Phys* **56** 89-105
[49] Swindell W, Simpson R G, Oleson J R, Chen C T and Grubbs E A 1983 Computed tomography with a linear accelerator with radiotherapy applications. *Med Phys* **10** 416-20
[50] Brahme A and Agren A K 1987 Optimal dose distribution for eradication of heterogeneous tumours. *Acta Oncol* **26** 377-85
[51] Mosleh-Shirazi M A, Swindell W and Evans P M 1998 Optimization of the scintillation detector in a combined 3D megavoltage CT scanner and portal imager [In Process Citation]. *Med Phys* **25** 1880-90
[52] Hesse B M, Spies L and Groh B A 1998 Tomotherapeutic portal imaging for radiation treatment verification . *Phys Med Biol* **43** 3607-16
[53] Jaffray D A, Drake D G, Moreau M, Martinez A A and Wong J W 1999 A radiographic and tomographic imaging system integrated into a medical linear accelerator for localization of bone and soft-tissue targets. *Int J Radiat Oncol Biol Phys* **45** 773-89

[54] Mutic S, Dempsey J F, Bosch W R, Low D A, Drzymala R E, Chao K S C et al. 2001 Multimodality image registration quality assurance for conformal three-dimensional treatment planning. *Int J Radiat Oncol Biol Phys* **51** 255-60

[55] Jaffray D A and Siewerdsen J.H 2000 Cone-beam computed tomography with a flat-panel imager: initial performance characterization. *Med Phys* **27** 1311-23

[56] Jaffray D A, Siewerdsen J H, Wong J W and Martinez A A 2002 Flat-panel cone-beam computed tomography for image-guided radiation therapy. *Int J Radiat Oncol Biol Phys* **53** 1337-49

[57] Mackie T R 2006 History of tomotherapy. *Phys Med Biol* **51** R427-R453

[58] Amols H I, Jaffray D A and Orton C G 2006 Point/counterpoint. Image-guided radiotherapy is being overvalued as a clinical tool in radiation oncology. *Med Phys* **33** 3583-6

[59] Jaffray D A 2007 Image-guided radiation therapy: from concept to practice. *Semin Radiat Oncol* **17** 243-4

[60] Moseley D J, White E A, Wiltshire K L, Rosewall T, Sharpe M B, Siewerdsen J H et al. 2007 Comparison of localization performance with implanted fiducial markers and cone-beam computed tomography for on-line image-guided radiotherapy of the prostate. *Int J Radiat Oncol Biol Phys* **67** 942-53

[61] Purdie T G, Bissonnette J P, Franks K, Bezjak A, Payne D, Sie F et al. 2007 Cone-beam computed tomography for on-line image guidance of lung stereotactic radiotherapy: localization, verification, and intrafraction tumor position. *Int J Radiat Oncol Biol Phys* **68** 243-52

[62] McBain C A, Henry A M, Sykes J, Amer A, Marchant T, Moore C M et al. 2006 X-ray volumetric imaging in image-guided radiotherapy: the new standard in on-treatment imaging. *Int J Radiat Oncol Biol Phys* **64** 625-34

[63] Woodford C, Yartsev S, Dar A R, Bauman G and Van Dyk J 2007 Adaptive radiotherapy planning on decreasing gross tumor volumes as seen on megavoltage computed tomography images. *Int J Radiat Oncol Biol Phys* **69** 1316-22

[64] Yartsev S, Kron T and Dyk J V 2007 Tomotherapy as a tool in image-guided radiation therapy (IGRT): current clinical experience and outcomes. *Biomed Imaging Interv J* **3** e17

[65] Brock K K, Dawson L A, Sharpe M B, Moseley D J and Jaffray D A 2006 Feasibility of a novel deformable image registration technique to facilitate classification, targeting, and monitoring of tumor and normal tissue. *Int J Radiat Oncol Biol Phys* **64** 1245-54

[66] Brock K K, Hawkins M, Eccles C, Moseley J L, Moseley D J, Jaffray D A et al. 2008 Improving image-guided target localization through deformable registration. *Acta Oncol* **47** 1279-85

[67] Craig T, Wong E, Bauman G, Battista J and Van Dyk J 2005 Impact of geometric uncertainties on evaluation treatment techniques for prostate cancer. *Int J Radiat Oncol Biol Phys* **62** 426-36

[68] Ahunbay E E, Peng C, Chen G P, Narayanan S, Yu C, Lawton C et al. 2008 An on-line replanning scheme for interfractional variations. *Med Phys* **35** 3607-15

[69] Amies C, Bani-Hashemi A, Celi J C, Grousset G, Ghelmansarai F, Hristov D et al. 2006 A multi-platform approach to image guided radiation therapy (IGRT). *Med Dosim* **31** 12-9

[70] Fox C, Romeijn H E, Lynch B, Men C, Aleman D M and Dempsey J F 2008 Comparative analysis of 60Co intensity-modulated radiation therapy. *Phys Med Biol* **53** 3175-88

[71] Kron T, Eyles D, Schreiner L J and Battista J J 2006 Magnetic resonance imaging for adaptive cobalt tomotherapy: A proposal. *J Med Phys* **31** 242-54

[72] Mackie T R, Holmes T, Reckwerdt P and Yang J 1995 Tomotherapy: Optimized Planning and Delivery of Radiation Therapy. *Int J Imaging Systems and Technology* **6** 43-55

[73] Welsh J S, Patel R R, Ritter M A, Harari P M, Mackie T R and Mehta M P 2002 Helical tomotherapy: an innovative technology and approach to radiation therapy. *Technol Cancer Res Treat* **1** 311-6

[74] Balter J M and Antonuk L E 2008 Quality assurance for kilo- and megavoltage in-room imaging and localization for off- and online setup error correction. *Int J Radiat Oncol Biol Phys* **71** S48-S52

[75] Bissonnette J P 2007 Quality assurance of image-guidance technologies. *Semin Radiat Oncol* **17** 278-86

[76] Bissonnette J P, Moseley D, White E, Sharpe M, Purdie T and Jaffray D A 2008 Quality assurance for the geometric accuracy of cone-beam CT guidance in radiation therapy. *Int J Radiat Oncol Biol Phys* **71** S57-S61

[77] Isambert A, Bonniaud G, Lavielle F, Malandain G and Lefkopoulos D 2008 A phantom study of the accuracy of CT, MR and PET image registrations with a block matching-based algorithm. *Cancer Radiother*

[78] Saw C B, Yang Y, Li F, Yue N J, Ding C, Komanduri K et al. 2007 Performance characteristics and quality assurance aspects of kilovoltage cone-beam CT on medical linear accelerator. *Med Dosim* **32** 80-5

[79] Sharpe M and Brock K K 2008 Quality assurance of serial 3D image registration, fusion, and segmentation. *Int J Radiat Oncol Biol Phys* **71** S33-S37

[80] Van Dyk J 2008 Quality assurance of radiation therapy planning systems: current status and remaining challenges. *Int J Radiat Oncol Biol Phys* **71** S23-S27

[81] Bissonnette J P, Moseley D J and Jaffray D A 2008 A quality assurance program for image quality of cone-beam CT guidance in radiation therapy. *Med Phys* **35** 1807-15

[82] Basran P S and Woo M K 2008 An analysis of tolerance levels in IMRT quality assurance procedures. *Med Phys* **35** 2300-7

[83] Both S, Alecu I M, Stan A R, Alecu M, Ciura A, Hansen J M et al. 2007 A study to establish reasonable action limits for patient-specific quality assurance in intensity-modulated radiation therapy. *J Appl Clin Med Phys* **8** 1-8

[84] Broggi S, Cattaneo G M, Molinelli S, Maggiulli E, Del Vecchio A, Longobardi B et al. 2008 Results of a two-year quality control program for a helical tomotherapy unit. *Radiother Oncol* **86** 231-41

[85] Howell R M, Smith I P and Jarrio C S 2008 Establishing action levels for EPID-based QA for IMRT. *J Appl Clin Med Phys* **9** 2721

[86] Jin H, Palta J, Suh T S and Kim S 2008 A generalized a priori dose uncertainty model of IMRT delivery. *Med Phys* **35** 982-96

[87] Marks L B, Light K L, Hubbs J L, Georgas D L, Jones E L, Wright M C et al. 2007 The impact of advanced technologies on treatment deviations in radiation treatment delivery. *Int J Radiat Oncol Biol Phys* **69** 1579-86

[88] Rangel A, Ploquin N, Kay I and Dunscombe P 2008 Evaluation of linear accelerator performance standards using an outcome oriented approach. *Med Phys* **35** 2513-8

[89] Hulick P R and Ascoli F A 2005 Quality assurance in radiation oncology. *J Am Coll Radiol* **2** 613-6

[90] Huq M S, Fraass B A, Dunscombe P B, Gibbons J P, Jr., Ibbott G S, Medin P M et al. 2008 A method for evaluating quality assurance needs in radiation therapy. *Int J Radiat Oncol Biol Phys* **71** S170-S173

[91] Palta J R, Liu C and Li J G 2008 Current external beam radiation therapy quality assurance guidance: does it meet the challenges of emerging image-guided technologies? *Int J Radiat Oncol Biol Phys* **71** S13-S17

[92] Williamson J F, Dunscombe P B, Sharpe M B, Thomadsen B R, Purdy J A and Deye J A 2008 Quality assurance needs for modern image-based radiotherapy: recommendations from 2007 interorganizational symposium on "quality assurance of radiation therapy: challenges of advanced technology". *Int J Radiat Oncol Biol Phys* **71** S2-12

[93] De Deene Y 2002 Gel dosimetry for the dose verification of intensity modulated radiotherapy treatments. *Z Med Phys* **12** 77-88

Journal of Physics: Conference Series **164** (2009) 012001 doi:10.1088/1742-6596/164/1/012001

[94] Gustavsson H, Karlsson A, BA[currency sign]ck SA, Olsson L E, Haraldsson P, EngstrA m P et al. 2003 MAGIC-type polymer gel for three-dimensional dosimetry: intensity-modulated radiation therapy verification. *Med Phys* **30** 1264-71

[95] Low D A, Gerber R L, Mutic S and Purdy J A 1998 Phantoms for IMRT dose distribution measurement and treatment verification. *Int J Radiat Oncol Biol Phys* **40** 1231-5

[96] Oldham M, Baustert I, Lord C, Smith T A, McJury M, Warrington A P et al. 1998 An investigation into the dosimetry of a nine-field tomotherapy irradiation using BANG-gel dosimetry. *Phys Med Biol* **43** 1113-32

[97] Scheib S G and Gianolini S 2002 Three-dimensional dose verification using BANG gel: a clinical example. *J Neurosurg* **97** 582-7

[98] Oldham M, Sterling D, Sakhalkar H, Molineau A, Adamovics J, and Ibbott G 2008 Evaluation of PRESAGE/optical-CT 3D Dosimetry for Commissioning a Linac for IMRT *Med Phys* **35** 2757-2757

[99] Oldham M, Sakhalkar H, Adamovics, Molineau A and Ibbott G 2008 The Feasibility of Comprehensive IMRT Verification using Novel 3D Dosimetry Techniques Compatible with the RPC Head and Neck Phantom *Int J Radiat Oncol Biol Phys* **72** S145-S145

[100] Babic S, Battista J and Jordan K 2008 Three-dimensional dose verification for intensity-modulated radiation therapy in the radiological physics centre head-and-neck phantom using optical computed tomography scans of ferrous xylenol-orange gel dosimeters. *Int J Radiat Oncol Biol Phys* **70** 1281-91

[101] Lopatiuk-Tirpak O, Langen KM, Meeks SL, Kupelian PA, Zeidan OA and Maryanski MJ 2008 Performance evaluation of an improved optical computed tomography gel dosimeter system for 3D dose verification of static and dynamic phantom deliveries *Med Phys* **35** 3847-3859

Journal of Physics: Conference Series **164** (2009) 012002 doi:10.1088/1742-6596/164/1/012002

Historical overview of the development of gel dosimetry: another personal perspective

Clive Baldock

Institute of Medical Physics, School of Physics, University of Sydney, NSW 2006, Australia

Email: c.baldock@physics.usyd.edu.au

1. Introduction

Over many years individuals have endeavoured to measure absorbed radiation dose distributions using gels. As long ago as 1950, the radiation-induced colour change in dyes was used to investigate radiation doses in gels [1]. Further, in 1957 depth doses of photons and electrons in agar gels were investigated using spectrophotometry [2]. Gel dosimetry today however, is founded mainly on the work of Gore *et al* who in 1984 [3] demonstrated that changes due to ionising radiation in Fricke dosimetry solutions [4], developed in the 1920's, could be measured using nuclear magnetic resonance (NMR).

2. Fricke Gel Dosimeters

Gore *et al* investigated [3] the nuclear magnetic resonance (NMR) relaxation properties of irradiated Fricke or ferrous sulphate dosimetry solutions [4] showing that radiation-induced changes, in which ferrous (Fe^{2+}) ions are converted to ferric (Fe^{3+}) ions, could be quantified using NMR relaxation measurements. In 1986 Appleby *et al* [5] reported that Fricke dosimetry solutions dispersed throughout a gel matrix could be used to obtain three-dimensional (3D) spatial dose information using magnetic resonance imaging (MRI). It was subsequently shown that irradiated Fricke-type gel dosimeters did not retain a spatially stable dose distribution due to ion diffusion within the irradiated dosimeters [6]. Fricke solutions with various gelling agents such as gelatine, agarose, sephadex and polyvinyl alcohol (PVA) were investigated along with chelating agents such as xylenol orange (XO) to reduce diffusion. Numerous authors subsequently published results of their work to inhibit the ion diffusion with limited success and summarised by Baldock in 2001 [7]. By the early 1990's the diffusion problem was considered to be a significant one in the advancement of gel dosimetry.

3. Polymer Gel Dosimeters

Polymer systems for the use of radiation dosimetry were first proposed as early as 1954, where Alexander *et al* [8] discussed the effects of ionising radiation on polymethylmethacrylate. Following this, Hoecker *et al* [9] in 1958 investigated the dosimetry of radiation-induced polymerisation in liquids, and in 1961 Boni [10] used polyacrylamide as a gamma dosimeter. Much later in 1991, Audet *et al* [11] reported changes in NMR transverse relaxation measurements of irradiated polyethelene oxide. In 1992, Kennan *et al* [12] reported on NMR longitudinal relaxation studies performed on an irradiated aqueous solution of N,N'-methylene-bis-acrylamide and agarose, which showed that the relaxation rates increased with absorbed dose.

In 1992 a new gel dosimetry formulation was proposed by Maryanski, which was based on the polymerisation of acrylamide and N,N'-methylene-bis-acrylamide (bis) monomers infused in an aqueous agarose matrix [13]. This system was given the acronym BANANA due to the use of the chemical components (bis, acrylamide, nitrous oxide and agarose) [14]. This type of gel dosimeter did not have the associated diffusion problem of Fricke gels and was shown to have a relatively stable post-irradiation dose distribution. The polymerisation reaction occurred by cross-linking of the monomers induced by the free radical products of water radiolysis. In 1994 the BANANA formulation was refined [15] by replacing agarose with gelatine and given the acronym BANG (bis, acrylamide, nitrogen and aqueous gelatine), the first of a series of new polymer gel formulations. In 1994 this formulation was patented [16] and became commercially available through MGS Research Inc. as BANG®. Subsequently, due to the naming of the commercial product, PAG [17] became the polymer gel dosimeter acronym of choice for most authors. Numerous authors subsequently published results of work investigating different compositions and formulations of polymer gel dosimeters which were summarised by Lepage *et al* [18].

Although polymer-type dosimeters did not have the diffusion limitations of Fricke-type gel dosimeters, there was another significant limitation in their use. Due to the nature of their free radical chemistry, polymer gel dosimeters were susceptible to atmospheric oxygen inhibition of the polymerisation processes. As a result, these gel dosimeters had to be manufactured in an oxygen-free environment [17,19], such as in a glove box pumped with nitrogen gas. Along with the use of potentially toxic chemicals [20], this was a significant limitation in the introduction of gel dosimetry into the clinic.

During this period a number of studies were undertaken to investigate the clinical applications of radiological tissue-equivalent [21,22,23] PAG-type polymer gel dosimeters using MRI [14,15,24,25,26]. De Deene et al [27] undertook an investigation into the overall accuracy of an anthropomorphic polymer gel dosimetry phantom for the verification of conformal radiotherapy treatments. It was established that significant issues relating to the accuracy of this dosimetry technique were a result of oxygen inhibition in the polymer gel and MRI imaging artefacts.

Authors continued to investigate clinical aspects of polymer gel dosimetry using MRI including conformal therapy, IMRT and IMAT [28-34], stereotactic radiosurgery [35-42], brachytherapy [43,44], low energy X-rays [46], high-LET and proton therapy [47-50], boron capture neutron therapy [51,52] and tissue inhomogeneities [29,31].

4. Normoxic Polymer Gel Dosimeters

A significant development in the field of gel dosimetry occurred when results of using an alternative polymer gel dosimeter formulation were published by Fong *et al* in 2001 [53]. This new type of polymer gel dosimeter, know as MAGIC gel, bound atmospheric oxygen in a metallo-organic complex thus removing the problem of oxygen inhibition and enabling polymer gels to be manufactured on the bench-top in the laboratory. This created what was to be known as a normoxic gel dosimeter, compared with the previous PAG formulation which subsequently became known as a hypoxic gel dosimeter. The MAGIC polymer gel formulation consisted of methacrylic acid, ascorbic acid, gelatine and copper. The principal behind the MAGIC gel is in the ascorbic acid oxygen scavenger. Ascorbic acid binds free oxygen contained within the aqueous gelatine matrix into metallo-organic complexes and this process is initiated by copper sulphate. It was subsequently shown by De Deene *et al* in 2002 that other antioxidants could be used in the manufacture of normoxic gels [54] including tetrakis (hydroxymethyl) phosphonium chloride, having first been suggested to Baldock by Billingham in 1996 [55]. Numerous authors subsequently published results of work investigating different compositions and formulations of normoxic polymer gel dosimeters and were recently summarised by Senden [56]. Recent work has also included the development of less toxic polymer gels [56].

With the introduction of normoxic gel dosimeters, MRI studies were undertaken to investigate their usefulness for IMRT [57], and radionuclide therapy [58-60].

5. Evaluation of Gel Dosimeters

Since the work of Gore *et al* in 1984 [3], the majority of investigations have been undertaken with MRI. However, in 1996 Gore et al and Maryanski et al demonstrated the potential of optical-CT as an alternative imaging technique to MRI for PAG-type polymer gel dosimeters [61,62]. This technique was further investigated by Oldham et al [63-65]. In 2000 Hilts et al demonstrated the use of X-ray CT to image PAG-type gels and subsequently used X-ray CT to investigate stereotactic dose distributions [66]. In 1998 Baldock demonstrated the use of variational spectroscopy to evaluate PAG-type polymer gel dosimeters [67,68] and in 2002 Mather et al demonstrated the use of ultrasound to image polymer gel dosimeters [69].

6. PRESAGE™ Dosimeters

A new class of polymer dosimeter, PRESAGE™ (Heuris Pharma, Skillman, NJ) [70] was proposed in 2003 and based on clear polyurethane combined with leuco-dye leucomalachite green. The components of the dosimeter, which was subsequently patented in 2006 [71], include an alkyl diisocyanate prepolymer, a hydroxyl reactive polyol along with a catalyst, which polymerises into optically clear polyurethane. Although not suitable for MRI evaluation, this radiation tissue-equivalent dosimeter [23] contains leuco dyes which have a maximum absorbance at a wavelength of 633 nm and are therefore suitable for evaluation with a He-Ne laser-based optical scanning system [72-74].

7. Other Non-radiotherapy Developments

A non-radiotherapy development in gel dosimetry was reported by Gore *et al* in 1997 using polymer gel dosimeters to develop an image quality test tool for MRI [75]. Hill reported the use of polymer gel dosimeters in measuring the CTDI (Computer Tomography Dose Index) on diagnostic x-ray CT scanners [76,77] indicating the potential of using polymer gel dosimeters for diagnostic dose levels.

8. Discussion and Conclusions

This paper is based on the previous review presented by the author at DOSGEL 2006 [78] and a more comprehensive review on polymer gel dosimetry which has been submitted for publication in *Physics in Medicine and Biology* [79]. Yves De Deene, Simon Doran, Geoff Ibbott, Andrew Jirasek, Martin Lepage, Kim McAuley, Mark Oldham and L John Schreiner are acknowledged as co-authors and co-contributors to the more comprehensive submitted review on polymer gel dosimetry.

An historical review of this nature cannot be comprehensive but only give an indication of the historical developments in the field of gel dosimetry. There are many papers which have made a significant contribution to the literature over the years which have not been mentioned here. This in no way diminishes the work of the authors of those papers. Indeed, as stated by Isaac Newton "if I have seen a little further it is by standing on the shoulders of giants". It is perhaps on the shoulders of the giants who have gone before that many of the developments in gel dosimetry have been achieved.

There have been a limited number of previous reviews on polymer gel dosimetry [80]. However, in a previous review on gel dosimetry [81] based on meta-analysis, potentially erroneous conclusions were drawn from a limited number of publications [82]. Readers are therefore encouraged to read the literature as well as the proceedings of the various DOSGEL conferences to comprehensively see how the discipline of gel dosimetry has developed over the years [83-86].

As was mentioned by the author in his previous paper [78], to date there is still no consensus on optimal gel formulations of gel dosimeters or the optimal evaluation techniques to be used. A result is that this form of 3D dosimetry is yet to be accepted for routine use in the clinic. Until then, there remains much research still to be undertaken.

9. References

[1] Day M J and Stein G 1950 Chemical effects of ionizing radiation in some gels *Nature* **166** 146–7

[2] Andrews H L, Murphy R E and LeBrun E J 1957 Gel dosimeter for depth dose measurements *Rev Sci Instrum* **28** 329–32

[3] Gore J C, Kang Y S and Schulz R J 1984 Measurement of radiation dose distributions by nuclear magnetic resonance (NMR) imaging *Phys Med Biol* **29** 1189–97

[4] Fricke H and Morse S 1927 The chemical action of roentgen rays on dilute ferrous sulphate solutions as a measure of radiation dose *Am J Roentgenol Radium Therapy Nucl Med* **18** 430–2

[5] Appleby A, Christman E A and Leghrouz A 1986 Imaging of spatial radiation dose distribution in agarose gels using magnetic resonance *Med Phys.* **14** 382-4

[6] Schulz R J, de Guzman A F, Nguyen D B and Gore J C 1990 Dose-response curves for Fricke-infused agarose gels as obtained by nuclear magnetic resonance *Phys Med Biol* **35** 1611-22

[7] Baldock C, Harris P J, Piercy A R, Healy B 2001 Experimental determination of the diffusion coefficient in two-dimensions in ferrous sulphate gels using the finite element method *Australas Phys Eng Sci Med* **24** 19-30

[8] Alexander P, Charlesby A and Ross M 1954 The degradation of solid polymethylmethacrylate by ionizing radiations *Proceedings of the Royal Society* **A223** 392

[9] Hoecker F E and Watkins I W 1958 Radiation polymerization dosimetry *Int J Appl Rad Isotop* **3** 31-35

[10] Boni A L 1961 A polyacrylamide gamma dosimeter *Radiation Research* **14** 374-80

[11] Audet C and Schreiner L J 1991 Radiation dosimetry by NMR relaxation time measurements of irradiated polymer solutions *Proc Intl Soc Mag Reson Med (New York)*

[12] Kennan R P, Maryanski M J, Zhong J and Gore J C 1992. Hydrodynamic effects and cross relaxation in cross linked polymer gels *Proc Intl Soc Mag Reson Med (New York)*

[13] Maryanski M J, Gore J C and Schulz R J 1992 3-D radiation dosimetry by MRI: solvent proton relaxation enhancement by radiation-controlled polymerisation and cross-linking in gels *Proc Intl Soc Mag Reson Med (New York)*

[14] Maryanski M J, Gore J C, Kennan R P and Schulz R J 1993 NMR relaxation enhancement in gels polymerized and cross-linked by ionizing radiation: a new approach to 3D dosimetry by MRI *Magn Reson Imaging* **11** 253-58

[15] Maryanski M J, Schulz R J, Ibbott G S, Gatenby J C, Xie J, Horton D and Gore J C 1994 Magnetic resonance imaging of radiation dose distributions using a polymer-gel dosimeter *Phys Med Biol* **39** 1437-55

[16] Maryanski M J, Gore J C and Schulz R 1994 Three-dimensional detection, dosimetry and imaging of an energy field by formation of a polymer in a gel *US Patent 5321357*

[17] Baldock C, Burford R P, Billingham N, Wagner G S, Patval S, Badawi R D and Keevil S F 1998 Experimental procedure for the manufacture and calibration of polyacrylamide gel (PAG) for magnetic resonance imaging (MRI) radiation dosimetry *Phys Med Biol* **43** 695-702

[18] Lepage M, Jayasekera M, Bäck S Å J, Baldock C 2001 Dose resolution optimization of polymer gel dosimeters using different monomers *Phys Med Biol* **46** 2665-80

[19] De Deene Y, De Wagter C, Van Duyse B, Derycke S, De Neve W and Achten E 1998 Three-dimensional dosimetry using polymer gel and magnetic resonance imaging applied to the verification of conformal radiation therapy in head-and-neck cancer *Radiotherapy and Oncology* **48** 283–291

[20] Baldock C and Watson S 1999 Risk assessment for the manufacture of radiation dosimetry polymer gels in *DOSGEL 1999 Proceedings of the 1st International Workshop on Radiation Therapy Gel Dosimetry* (Lexington, USA) Eds L J Schreiner and C Audet

[21] Keall P, Baldock C, 1999. A theoretical study of the radiological properties and water equivalence of three types of gels used for radiation dosimetry *Australas Phys Eng Sci Med* **22** 85-91

[22] Venning AJ, Nitschke KN, Keall PJ, Baldock C, 2005. Radiological properties of normoxic polymer gel dosimeters *Med Phys* **32** 1047-1053

[23] Brown S, Venning A, De Deene Y, Vial P, Oliver L, Adamovics J and Baldock C 2008 Radiological properties of the PRESAGE and PAGAT polymer dosimeters *Applied Radiation and Isotopes* (in press)

[24] Ibbott G S, Maryanski M J, Eastman P, Holcomb S D, Zhang Y, Avison R G, Sanders M and Gore J C 1997 Three-dimensional visualization and measurement of conformal dose distributions using magnetic resonance imaging of BANG polymer gel dosimeters *Int J Radiat Oncol Biol Phys* **38** 1097-103

[25] Oldham M, Baustert I, Lord C, Smith T A D, McJury M, Warrington A P, Leach M O and Webb S 1998a An investigation into the dosimetry of a nine-field tomotherapy irradiation using BANG-gel dosimetry *Phys Med Biol* **43** 1113–32

[26] Low D A, Harms W B, Mutic S and Purdy J A 1998 A technique for the quantitative evaluation of dose distributions *Med Phys* **25** 656-61

[27] De Deene Y, De Wagter C, Van Duyse B, Derycke S, Mersseman B, De Gersem W, Voet T, Achten E and De Neve W 2000 Validation of MR-based polymer gel dosimetry as a preclinical three-dimensional verification tool in conformal radiotherapy *Magn Reson Med* **43** 116–25

[28] Cosgrove V P, Murphy P S, McJury M, Adams E J, Warrington A P, Leach M O and Webb S 2000 The reproducibility of polyacrylamide gel dosimetry applied to stereotactic conformal radiotherapy *Phys Med Biol* **45** 1195-210

[29] Vergote K, De Deene Y, Claus F, De Gersem W, Van Duyse B, Paelinck L, Achten E, De Neve W, De Wagter C 2003 Application of monomer/polymer gel dosimetry to study the effects of tissue inhomogeneities on intensity-modulated radiation therapy (IMRT) dose distributions *Radiotherapy and Oncology* **67** 119-28

[30] Duthoy W, De Gersem W, Vergote K, Coghe M, Boterberg T, De Deene Y, De Wagter C, Van Belle S and De Neve W 2003 Whole abdominopelvic radiotherapy (WAPRT) using intensity-modulated arc therapy (IMAT): First clinical experience *Int J Radiation Oncology Biol Phys* **57** 1019-32

[31] Love P A, Evans P M, Leach M O and Webb S 2003 Polymer gel measurement of dose homogeneity in the breast: comparing MLC intensity modulation with standard wedged delivery *Phys Med Biol* **48** 1065-74

[32] Vergote K, De Deene Y, Duthoy W, De Gersem W, De Neve W, Achten E 2004 Validation and application of polymer gel dosimetry for the dose verification of an intensity-modulated arc therapy (IMAT) treatment *Phys Med Biol* **49** 287-305

[33] Duthoy W, De Gersem W, Vergote K, Boterberg T, Derie C, Smeets P, De Wagter C and De Neve W 2004 Clinical implementation of intensity-modulated arc therapy (IMAT) for rectal cancer *Int J Radiation Oncology Biol Phys* **60** 794-806

[34] Sandilos P, Angelopoulos A, Baras P, Dardoufas K, Karaiskos P, Kipouros P, Kozicki M, Rosiak J M, Sakelliou L, Seimenis I and Vlahos L 2004 Dose verification in clinical IMRT prostate incidents *Int J Radiation Oncology Biol Phys* **59** 1540-7

[35] Ertl A, Berg A, Zehetmayer M and Frigo P 2000 High-resolution dose profile studies based on MR imaging with polymer BANG gels in stereotactic radiation techniques *Magn Reson Imaging* **18** 343-349

[36] Grebe G, Pfaender M, Roll M and Luedemann L 2001 Dynamic arc radiosurgery and radiotherapy: Commissioning and verification of dose distributions *Int J Radiat Oncol Biol Phys* **49** 1451-60

[37] Pappas E, Seimenis I, Angelopoulos A, Georgolopoulou P, Kamariotaki Paparigopoulou M, Maris T, Sakelliou L, Sandilos P and Vlachos L 2001 Narrow stereotactic beam profile measurements using N-vinylpyrrolidone based polymer gels and magnetic resonance imaging *Phys Med Biol* **46** 783-97

[38] Audet C, Hilts M, Jirasek A and Duzenli C 2002 CT gel dosimetry technique: Comparison of a planned and measured 3D stereotactic dose volume *J Appl Clin Med Phys* **3** 110-8

[39] Novotny J Jr, Dvorak P, Spevacek V, Tintera J, Novotny J, Cechak T and Liscak R 2002 Quality control of the stereotactic radiosurgery procedure with the polymer-gel dosimetry *Radiother Oncol* 63 223-30

[40] Scheib S G and Gianolini S 2002 Three-dimensional dose verification using BANG gel: a clinical example *J Neurosurg* **97** 582-7

[41] Watanabe Y, Perera G M and Mooij R B 2002 Image distortion in MRI-based polymer gel dosimetry of Gamma Knife stereotactic radiosurgery systems *Med Phys* **29** 797-802

[42] Karaiskos P, Petrokokkinos L, Tatsis E, Angeloupolos A, Baras P, Kozicki M, Papagiannis P, Rosiak J M, Sakelliou L, Sandilos P and Vlachos L 2005 Dose verification of single shot gamma knife applications using VIPAR polymer gel and MRI *Phys Med Biol* **50** 1235-50

[43] Farajollahi A R, Bonnett D E, Ratcliffe A J, Aukett R J and Mills J A 1999 An investigation into the use of polymer gel dosimetry in low dose rate brachytherapy *Br J Radiol* **72** 1085–92

[44] Wuu C-S , Schiff P, Maryanski MJ, Liu T, Borzillary S, and Weinberger J 2003 Dosimetry study of

[45] Re-188 liquid balloon for intravascular brachytherapy using polymer gel dosimeters and laser-beam optical CT scanner *Med Phys* **30** 132-7

[46] Boudou C, Briston M C, Corde S, Adam J F, Ferrero C, Esteve F and Elleaume H 2004 Synchrotron stereotactic radiotherapy: dosimetry by Fricke gel and Monte Carlo simulations *Phys Med Biol* **49** 5135-44

[47] Ramm U, Weber U, Bock M, Kramer M, Bankamp A, Damrau M, Thilmann C, Bottcher H D, Schad L R, and Kraft G 2000 Three-dimensional BANG gel dosimetry in conformal carbon ion radiotherapy *Phys Med Biol* **45** N95-N102

[48] Jirasek A I and Duzenli C 2002 Relative effectiveness of polyacrylamide gel dosimeters applied to proton beams: Fourier transform Raman observations and track structure calculations *Med Phys* **29** 569-77

[49] Heufelder J, Stiefel S, Pfaender M, Ludemann L, Grebe G and Heese J 2003 Use of BANG polymer gel for dose measurements in a 68 MeV proton beam *Med Phys* **30** 1235-40

[50] Gustavsson H, Back S A J, Medin J, Grusell E and Olsson L E 2004 Linear energy transfer dependence of a normoxic polymer gel dosimeter investigated using proton beam absorbed dose measurements *Phys Med Biol* **49** 3847-55

[51] Farajollahi A R, Bonnett D E, Tattam D and Green S 2000 The potential use of polymer gel dosimetry in boron neutron capture therapy *Phys Med Biol* **45** N9–N14

[52] Gambarini G, Colli V, Gay S, Petrovich C, Pirola L and Rosi G 2004 In-phantom imaging of all dose components in boron neutron capture therapy by means of gel dosimeters *Applied Radiation and Isotopes* **61** 759–763

[53] Fong P M, Keil D C, Does M D and Gore J C 2001 Polymer gels for magnetic resonance imaging of radiation dose distributions at normal room atmosphere *Phys Med Biol* **46** 3105–13

[54] De Deene Y, Hurley C, Venning A, Mather M, Healy B, Whittaker A, Baldock C 2002 A basic study of some normoxic polymer gel dosimeters *Phys Med Biol* **47** 3441–63

[55] Billingham N, 1996. Personal communication.

[56] Senden R J, De Jean P, McAuley K B and Schreiner L J 2006 Polymer gel dosimeters with reduced toxicity: a preliminary investigation of the NMR and optical dose–response using different monomers *Phys Med Biol* **51** 3301–14

[57] Gustavsson H, Karlsson A, Back S A J, Olsson L E, Haraldsson P, EngstrÄom P and Nystrom H 2003 MAGIC-type polymer gel for three-dimensional dosimetry: Intensity-modulated radiation therapy verification *Med Phys* **30** 1264-71

[58] Courbon F, Love P, Chittenden S, Flux G, Ravel P, Cook G 2006 Preparation and use of I-131 MAGIC gel as a dosimeter for targeted radionuclide therapy *Cancer Biotherapy & Radiopharmaceuticals* **21** 427-36

[59] Gear J I, Flux G D, Charles-Edwards E, Partridge M, Cook G, Ott R J 2006 The application of polymer gel dosimeters to dosimetry for targeted radionuclide therapy *Phys. Med. Biol.* ***51*** 3503–16

[60] Gear J I, Charles-Edwards E, Partridge M, Flux G D 2007 A quality-control method for SPECT-based dosimetry in targeted radionuclide therapy *Cancer Biotherapy & Radiopharmaceuticals* **22** 166-74

[61] Gore J C, Ranade M, Maryanski M J, Schulz R J 1996 Radiation dose distributions in three dimensions from tomographic optical density scanning of polymer gels: I. Development of an optical scanner *Phys Med Biol* **41** 2695-704
[62] Maryanski M J, Zastavker Y Z and Gore J C 1996 Radiation dose distributions in three dimensions from tomographic optical density scanning of polymer gels: II. Optical properties of the BANG polymer gel *Phys Med Biol* **41** 2705-17
[63] Oldham M, Siewerdsen J H, Shetty A and Jaffray D A 2001 High resolution gel-dosimetry by optical-CT and MR scanning *Med Phys* **28** 1436-45
[64] Oldham M, Siewerdsen J H, Kumar S, Wong J And Jaffray D A 2003 Optical-CT gel dosimetry I: Basic investigations *Med Phys* **30** 623-34
[65] Oldham M and Kim L 2004 Optical CT-gel dosimetry: II. Optical artifacts and geometric distortion *Med Phys* **31** 1093-104
[66] Hilts M, Audet C, Duzenli C and Jirasek A 2000 Polymer gel dosimetry using x-ray computed tomography: A feasibility study *Phys Med Biol* **45** 2559-71
[67] Baldock C, Rintoul L, Keevil S F, Pope J M and George G A 1998 Fourier transform Raman spectroscopy of polyacrylamide gel (PAGs) for radiation dosimetry *Phys Med Biol* **43** 3617-27
[68] Rintoul L, Lepage M, Baldock C 2003 Radiation dose distribution in polymer gels by Raman spectroscopy *Appl Spectroscopy* **57** 51-57
[69] Mather M L, De Deene Y, Whitakker A K, Simon G P, Rutgers R and Baldock C 2002 Investigation of ultrasonic properties of PAG and MAGIC polymer gel dosimeters *Phys Med Biol* **47** 4397-409
[70] Adamovics J and Maryanski M 2003 New 3D radiochromic solid polymer dosimeter from leuco dyes and a transparent polymeric matrix *Med Phys* **30** 1349
[71] Adamovics J 2006 Three-dimensional dosimeter for penetrating radiation and method of use *US Patent 7098463*
[72] Guo P, Adamovics J and Oldham M 2006 Characterization of a new radiochromic three-dimensional dosimeter *Med Phys* **33** 1338-1345
[73] Guo P, Adamovics J and Oldham M 2006 A practical three-dimensional dosimetry system for radiation therapy *Med. Phys.* **33** 3962-3972
[74] Oldham M, Sakhalkar M, Guo P and Adamovics J 2008 An investigation of the accuracy of an IMRT dose distribution using two- and three-dimensional dosimetry techniques *Med Phys* **35** 2072-2080
[75] Gore J C Maryanski M J and Schulz 1997 Test objects for MRI quality assurance based on polymer gels *Med Phys* **34** 1405-1408
[76] Hill B, Venning A J, Baldock C 2005 A preliminary study of the novel application of normoxic polymer gel dosimeters for the measurement of CTDI on diagnostic x-ray CT scanners *Med Phys* **32** 1589-97
[77] Hill B, Venning A J, Baldock C 2008 Polymer gel dosimetry on a multislice computed tomography scanner: Effect of changing parameters on CTDI *Physica Medica* **24** 149-158
[78] Baldock C 2006 Historical overview of the development of gel dosimetry: a personal perspective *Journal of Physics: Conference Series* **56** 14–22
[79] Baldock C, De Deene Y, Doran S, Ibbott G, Jirasek A, Lepage M, McAuley K B, Oldham M and Schreiner L J 2008 Topical Review: Polymer gel dosimetry *Phys Med Biol* (Submitted)
[80] McJury M, Oldham M, Cosgrove V P, Murphy P S, Doran S, Leach M O and Webb S 2000 Radiation dosimetry using polymer gels: methods and applications. *Br J Radiology* **73** 919-29
[81] MacDougall N D, Pitchford WG and Smith M A 2002 A systematic review of the precision and accuracy of dose measurements in photon radiotherapy using polymer and Fricke MRI gel dosimetry *Phys Med Biol* **47** R107-21
[82] De Deene Y, De Wagter C and Baldock C 2003 Comment on `A systematic review of the precision and accuracy of dose measurements in photon radiotherapy using polymer and Fricke MRI gel dosimetry' *Phys Med Biol* **48** L15-L18
[83] *DOSGEL 1999 Proceedings of the 1st International Workshop on Radiation Therapy Gel Dosimetry* (Lexington, USA) Eds L J Schreiner and C Audet
[84] *DOSGEL 2001 Proceedings of the 2nd International Conference on Radiotherapy Gel Dosimetry* (Brisbane, Australia) Eds C Baldock and Y De Deene

[85] *DOSGEL 2004 Proceedings of the 3rd International Conference on Radiotherapy Gel Dosimetry* (Ghent, Belgium) Eds C Baldock and Y De Deene

[86] *DOSGEL 2006 Proceedings of the 4th International Conference on Radiotherapy Gel Dosimetry* (Sherbrooke, Canada) Eds M Lepage, A Jirasek and L J Schreiner

Journal of Physics: Conference Series **164** (2009) 012003 doi:10.1088/1742-6596/164/1/012003

How Does the Chemistry of Polymer Gel Dosimeters Affect their Performance?

A. Jirasek*, K. B. McAuley and M. Lepage*****

*Department of Physics and Astronomy, University of Victoria, Victoria BC Canada V8W 3P6

**Department of Chemical Engineering, Queen's University, Kingston ON Canada K7L 3N6

***Centre d'imagerie moléculaire de Sherbrooke, Université de Sherbrooke, 3001 12e Avenue Nord, Sherbrooke QC Canada J1H 5N4

jirasek@uvic.ca, kim.mcauley@chee.queensu.ca , martin.lepage@usherbrooke.ca

Abstract. This review article describes the primary chemical processes affecting polymerization in polymer-based radiosensitive gel dosimeters. Furthermore, recently studied environmental factors and gelation processes affecting polymerization and gel dose response are discussed. Finally, the predominant physical factors affecting dose-response sensitivity for each of the three imaging modalities (MRI, Optical CT, x-ray CT) are outlined. Emphasis is concentrated on recent literature in this area, as prior literature was summarized in review articles submitted to DOSGEL 2006 in Sherbrooke (J. Conf. Series, 56, 2006). Discussions from the recent DOSGEL conference (2008) are incorporated into this review.

1. Introduction

This review describes the chemistry of polymer gel dosimeters that use free radical polymerization as the means to capture and store 3-D radiation dose information (e.g., MAG [1], PAG [2], NIPAM [3], and VIPAR [4] dosimeters). Dosimeters that rely on other types of chemical reactions (e.g., Fricke gel [5] and PRESAGE [6] dosimeters) are not included. Review articles from DOSGEL 2006 [7] summarize relevant information prior to 2006, so the focus of the current article is on the most recent literature, including some new results presented at DOSGEL 2008. We begin with some recent insights into the influence of polymerization mechanisms on dose response behavior, followed by a description of new polymer gel dosimeter recipes and additives. The influence of environmental factors on dosimeter response is described, especially the importance of temperature history during gel manufacture. Recent knowledge about the influence of oxygen and other impurities on dosimeter response is also described, along with the influence of THPC (tetrakis hydroxymethylphosphonium chloride) and gelatin. Finally, we discuss the physical phenomena that influence dose-response measurements using three read-out modalities: MRI, optical CT and x-ray CT, and we conclude that different gel recipes are suitable for different modalities.

2. Polymerization Chemistry and Dose-Rate Dependence

Table 1 summarizes the basic chemical reactions that occur in impurity-free polymer gel dosimeters [8-10]. In reaction a), several different types of free radicals are generated by radiolysis of water. These free radicals can react with the vinyl group (carbon-carbon double bond) on monomer molecules. Successive propagation reactions (reaction b) result in the formation of polymer and the release of heat. Many dosimeters (e.g., PAG, NIPAM and VIPAR) contain N,N'-methylene bisacrylamide (Bis), which has two vinyl groups. Reaction c) shows propagation with one of the vinyl groups on a Bis molecule, leaving the remaining vinyl group available for a subsequent reaction. The second vinyl group can be consumed by cyclization (reaction d) or crosslinking (reaction e). Crosslinking is important in PAG, NIPAM and VIPAR dosimeters because crosslinked polymer chains are insoluble in water, and they precipitate to form microgels [11] that influence the NMR, optical and x-ray properties of the gel. PAG dosimeters that have high concentrations of Bis (up to its solubility limit) tend to have higher dose sensitivities than dosimeters with lower crosslinker levels [12, 13]. All current polymer gel dosimeter recipes use Bis [3] as a crosslinker, except for MAG dosimeters [1] and a new fluorescent dosimeter described in Section 3 below [14]. MAG dosimeter recipes are usually crosslinker-free because polymethacrylic acid precipitates to form insoluble particles even when no crosslinker is present. As a result, reactions c), d) and e) are not important in MAG dosimeters. De Deene et al. [15] manufactured and irradiated MAG dosimeters in which a portion of the methacrylic acid monomer was replaced by Bis to determine whether the addition of crosslinker could enhance the dose sensitivity. They determined that the transverse relaxation rate (R_2) dose sensitivity decreased with increasing Bis concentration, presumably because crosslinking is not required to induce precipitation and Bis is less reactive than methacrylic acid so that less polymer formed per Gy of absorbed radiation.

Table 1. Cartoon showing free radical polymerization mechanism in polymer gel dosimeters.

a) Generation of primary radicals by radiolysis	$H_2O + Radiation \rightarrow H_2, H_2O_2, e^-_{aq}, H^\bullet, OH^\bullet$
b) Propagation with monomer	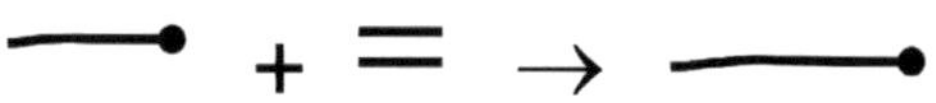
c) Propagation with crosslinker	
d) Primary cyclization	
e) Crosslinking	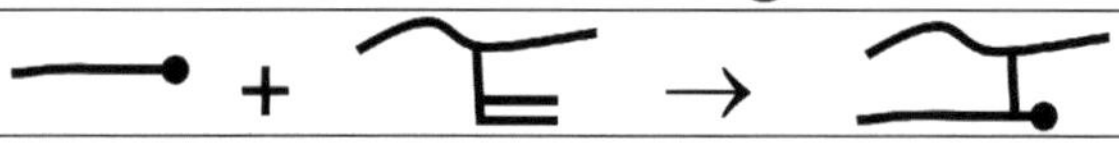
f) Chain transfer to monomer	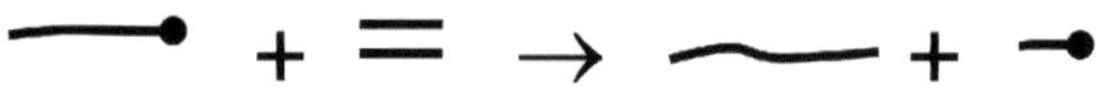
g) Termination	
h) Chain transfer to gelatin	
i) Propagation of gelatin-centred radicals	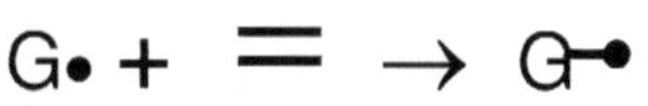
j) Termination with primary radicals	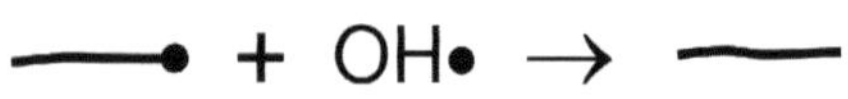

2.1. Dominant Chain-Stopping Mechanisms and their Influence on Dose Integration

De Deene et al. [16] and Karlsson et al. [19] observed that the dose response of MAG dosimeters depends on the dose rate. MAG dosimeters have significantly lower dose sensitivities when the radiation dose is delivered over a short period of time than when the same total dose is delivered over a longer period of time. In PAG and NIPAM dosimeters, however, the dose response is much less dose-rate dependent [3,15,18]. The reaction mechanism in Table 1 can explain this difference in dose-rate behaviour, and also the influence of gelatin on the dose sensitivity of PAG dosimeters [8].

In all polymer gel dosimeters, each polymer chain continues to grow via propagation and crosslinking reactions b), c) and e) until the radical is consumed via a chain-transfer reaction (e.g., reaction f) or by a bimolecular termination reaction (e.g. reaction f). Note that the radicals on crosslinked polymer molecules can react with pendant vinyl groups on other polymer molecules to form very large crosslinked molecules within the precipitated microgel particles.. Chain-transfer reactions consume one radical and generate another; termination reactions consume two radicals. Reaction g) shows a termination reaction between two growing polymeric radicals, but any two radicals that encounter each other can readily terminate. For example, termination between polymeric radicals and primary radicals generated by reaction a) also occurs (j). Reaction h) shows a chain-transfer reaction in which a growing polymeric radical abstracts a proton from a strand of gelatin, creating a dead polymer chain and a free radical on the gelatin. This gelatin radical can either terminate with another radical, or it can react with monomer (reaction i) or crosslinker to initiate further propagation. If reaction i) is slow because the gelatin radicals are relatively stable compared to polymeric radicals then many gelatin radicals terminate rather than propagate, thereby reducing the polymerization rate. Fuxman et al. [8,9] successfully used reactions h) and i) to explain the decrease in dose sensitivity with increasing gelatin levels that has been observed experimentally in PAG dosimeters (Figure 1) [15,16]. This mechanism can also explain [10] the high sensitivity to dose rate of MAG dosimeters and the low sensitivity to dose-rate observed in PAG [15,18] and NIPAM [3]

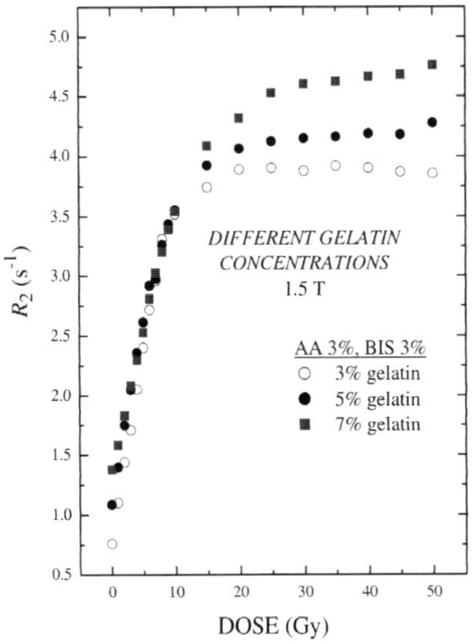

Figure 1: Dependence of the transverse relaxation rate (R_2) on the absorbed dose for anoxic polymer gel dosimeters with a fixed content of acrylamide (AA) and N,N'-methylenebisacrylamide (BIS) and a varying gelatin content. The initial slope is large and, the initial value of R_2 is lower for lower gelatin concentration. Data from [39].

dosimeters. Since polymeric radicals may react with gelatin, then it is also plausible that water free radicals may also react with gelatin, leaving fewer radicals to initiate polymerization.

In dosimeters like MAG and the new fluorescent dosimeter [14] where the dominant chain-stopping mechanism is termination between two radicals (e.g., reaction g), the concentration of the radicals in the solution has a large influence on average amount of polymer that forms. When radicals are generated quickly, the overall radical concentration is high and the termination rate is very high because the rate of termination is proportional to the square of the radical concentration. When the same number of radicals is generated over a longer time period, the radical concentration remains low and the termination rate is low, so that more polymer is generated per radical.

In PAG and NIPAM dosimeters, where gelatin radicals react very slowly with the monomer, the dominant chain-stopping mechanism is reaction h) rather than reaction g). Since reaction h) involves only a single radical, the rate of chain death is proportional to the radical concentration. Since the rates of propagation and chain-stopping reactions are both first-order in the radical concentration, the average amount of polymer generated by each radical is nearly independent of the radical concentration. As a result, there is little influence of dose rate on the total amount of polymer formed, making PAG and NIPAM dosimeters better integrating dosimeters than MAG.

2.2. The Influence of Gelatin on Dose Sensitivity

One peculiarity of MAG dosimeters that is not explained by the mechanism in Table 1 is that the dose sensitivity of these dosimeters increases rather than decreases with the amount of gelatin used in the recipe [15]. If reactions h) and i) are not important in MAG dosimeters, because either h) does not occur appreciably or because reaction i) is fast, then it is straightforward to predict that gelatin should have little or no effect on the rate of polymerization. However, it is not possible to predict an increase in the amount of polymer formed with increasing gelatin concentration from the reaction mechanism in Table 1. At DOSGEL 2006, McAuley [10] suggested that the gelatin might be causing changes in the properties of the aqueous solution, thereby helping more polymethacrylic acid to precipitate from solution when more gelatin is present. Her suggestion was based on previous studies [19-21], which showed that uncrosslinked polymethacrylic acid is sometimes soluble and sometimes insoluble in aqueous solutions, depending on the pH, ionic strength and temperature of the solution, and that solution properties also influence the molecular weight of the polymer. Gelatin could be influencing the size and number of polymethacrylic acid particles that precipitate, thereby influencing the dose sensitivity. Since DOSGEL 2006, it has become clear that this explanation is not sufficient to explain the influence of gelatin on dose sensitivity in MAG dosimeters. Recently, Hayashi et al. [22] used temperature probes in MAG dosimeters to monitor the rise in temperature associated with the heat released by the propagation reactions [23]. They showed that when no gelatin is present, there is no temperature rise and hence no appreciable polymerization. They also showed that there is an increase in the temperature rise per unit radiation dose when the amount of gelatin in the recipe increases. These experiments show definitively that gelatin influences the rate of polymerization rather than just the relative amounts of polymer that stay in solution and that precipitate.

Thorough examination of the free-radical polymerization literature shows that similar "trigger" phenomena have been observed previously in the polymerization of methacrylic acid and also acrylic acid in aqueous solutions. In 1968, Bamford and Shiiki [24] studied the free-radical polymerization of acrylic acid in the presence (and absence) of polyethyleneimine at 25 °C. When no polyethyleneimine was dissolved in the solution, the rate of polymerization was effectively zero. However, substantial polyacrylic acid formation was observed when polyethyleneimine was present in solution. The polymerization rate increased with increasing concentrations of polyethyleneimine until the molar concentration of base groups from the polyethyleneimine was approximately equal to the acrylic acid concentration in the solution. Bamford and Shiiki suggested that the polyethyleneimine acts as a "template"that complexes with the acrylic acid molecules, enabling their polymerization at a temperature where polymerization would not occur without the template. A review [25] of more-recent literature reveals that acrylic acid and methacrylic acid undergo template polymerization in the presence of several different water-soluble polymers containing basic groups, including poly(vinyl pyrrolidone) and poly(4-vinyl pyridine). The template polymerization of methacrylic acid using gelatin has not been previously reported in the literature, but the observations of DeDeene et al. [15] and Hayashi et al. [22] are consistent with this type of mechanism.

We already mentioned that OH• radicals can interact with the growing polymer chain and can cause termination of the polymerization reaction. The polymer formed by methacrylic acid monomers in MAG is more water-soluble than polymer containing Bis in PAG. Experiments also reveal more polymer is formed per unit dose from methacrylic acid, which suggests the methacryilic acid polymer radical is more reactive than its acrylamide polymer radical counterpart. This is consistent with the observation of a higher temperature increase in MAG dosimeters than in PAG dosimeters for an equivalent initial concentration of monomers and for the same irradiation dose. As a consequence, the methacrylic acid polymer radicals are more likely to react with additional OH• radicals. When adding gelatin, the number of OH• radicals becomes smaller, thus decreasing the termination of polymerization of methacrylic acid polymer radicals. This mechanism could also explain the dose-rate dependence of MAG and the observation that little or no MAG polymer is formed in the absence of gelatin. Clearly, more work is required to determine the mechanism by which gelatin affects the dose dependence in PAG and MAG dosimeters.

3. New Dosimeter Recipes

One new type of polymer gel dosimeter [14] was reported at DOSGEL 2008, and several new dosimeter additives, such as formaldehyde, have been studied since DOSGEL 2008 [27]. The new dosimeter [14] uses methylmethacrylate as the main monomer and small quantities of a fluorogenic probe species as a comonomer. When the comonomer molecule is incorporated into a polymer change, it changes from being very weakly fluorescent to highly fluorescent, so that the dose distribution can be determined using fluorometric dose imaging. This dosimeter, which does not include water, is not tissue-equivalent. A more serious issue is that this dosimeter is highly dose-rate dependent because the free radicals terminate solely via biomolecular termination.

A new dosimeter additive that has been reported is formaldehyde, which is used to increase the melting temperature and gel strength of MAG gels [26]. This additive also increases the dose sensitivity, but the mechanism for this sensitivity increase is not known.

Co-solvents are another type of additive that has been investigated for improving the effectiveness of gels that use Bis as a crosslinker [27]. Glycerol and isopropanol are two co-solvents that can be added to dosimeter recipes so that higher levels of Bis can be dissolved, leading to increased dose sensitivity due to increased crosslink density in the polymer. The motivation for trying to increase the dose sensitivity is so that new recipes can be developed that are suitable for x-ray CT readout.

4. Environmental Factors

A number of environmental factors during polymer gel manufacture affect the resultant dose response of the gels.

4.1. Temperature

DeDeene et al [28] reported on the effects of temperature history on polymer gel manufacture. They manufactured a series of gels and allowed them to cool under a range of conditions, including gels cooled in a fridge (with and without a water bath) and gels cooled under ambient conditions (see figure 2a). Figure 2b illustrates the cooling rate and time required to reach thermal equilibrium for the three different classes of gel cooling. Figure 3 illustrates the resulting variation in polymer gel response for nPAG, nMAG, and VIPAR gels. As can be seen, nPAG gels are less susceptible (7% variation at 10 Gy) to rates of gel cooling during the solidification process than are nMAG gels (67% at 10 Gy).

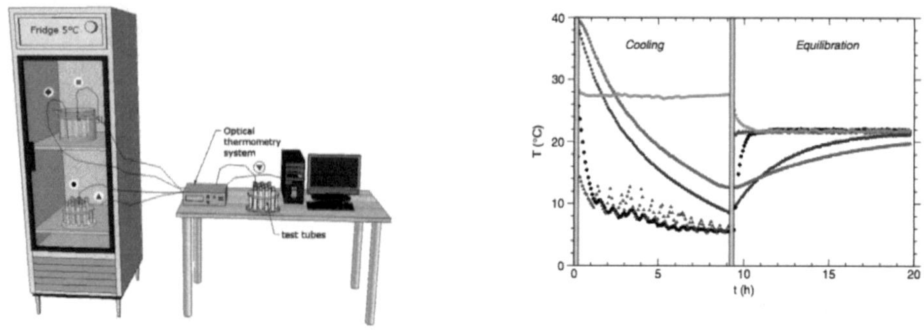

(a) (b)

Figure 2: (a) Gels manufactured and cooled using either (i) ambient room temperature, (ii) a fridge, or (iii) a fridge with gels immersed in a water bath. (b) Cooling rates and time required to cool gels in (a) to equilibrium [28].

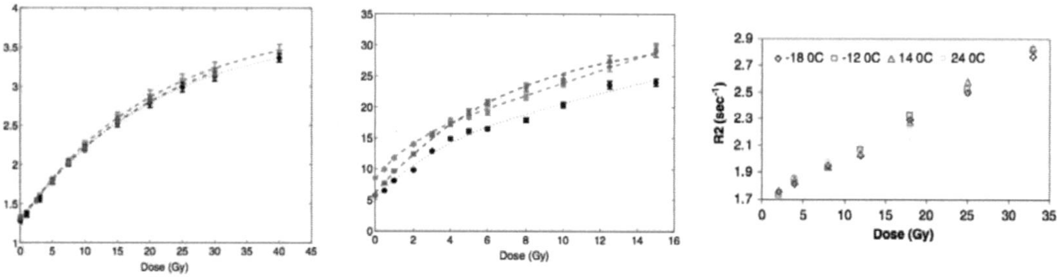

(a) (b) (c)

Figure 3: Variation in polymer gel dose response for gels with varying temperature history. (a) nPAG, (b) nMAG, and (c) VIPAR polymer gels. Temperature history of VIPAR gels is implicit, as in this case gels were cooled to the temperatures stated in the plot [28,29].

Polymer gel cooling rates are affected not only by their surrounding atmosphere, as demonstrated above, but also by the size of the gel. Figure 4 illustrates the variations in cooling rate within different regions of the gel for large (4L) gel volumes. The effect is observable irrespective of the presence of a water bath, although in this case the effect is mitigated. The discontinuity present in the cooling rate is due to the phase transition of the gel, hence affecting the rate of flow of convection currents. From figure 4 it can be seen that care must be taken in cooling large volumes of gel, with attempts made to minimize any differences in cooling rate in different regions of the gel volume.

DeDeene et al. [28] further investigated the effects of cooling rate on the accuracy of polymer gel dosimetry. They used a simple irradiation (depth dose) to compare ion chamber and gel measurements. Using 4 batches of gel they were able to establish absolute errors associated with gel measurements of the depth dose. Figure 4 illustrates the results and indicates that a maximum average dose difference of 0.5 Gy exists between gel and ion chamber measurements. Results indicate that, despite the given uncertainty in gel response due to different manufacture cooling rates, accurate dosimetry can still be performed under controlled conditions.

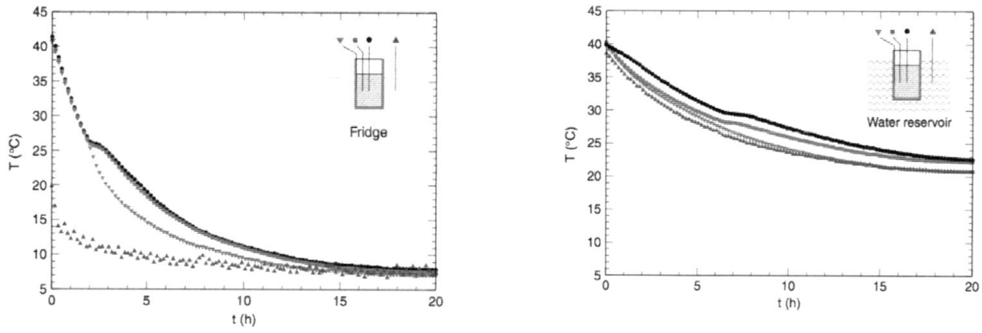

(a) (b)

Figure 4: Cooling rates as a function of position within the gel volume of large (4L) volumes of gel. Gels cooled in fridge (a) without and (b) with water bath [28].

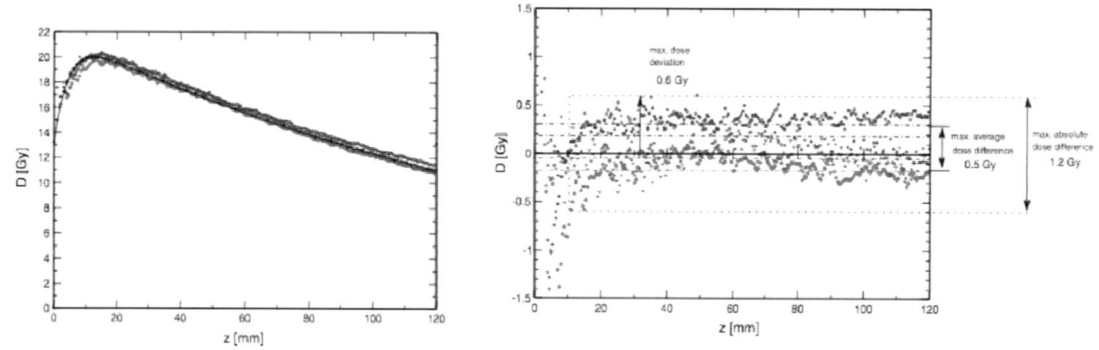

(a) (b)

Figure 5: (a) Depth dose cure measured using 4 batches of polymer gel and an ion chamber. (b) Dose difference between gel and ion chamber measurements for the curves in (a) [28].

5. Impurities in the Manufacture Process

The purity of the constituents used in polymer gel manufacture can affect, for example, the rate of oxygen scavenging when an antioxidant is used to de-oxygenate the gel [30]. A simple example is illustrated in figure 6. The rate of oxygen scavenging in tap water is significantly slower than the scavenging rate in deionized and purified water (figure 6a), illustrating the advantage of using deionized water in the manufacture process. Even with deionized water, the total time to scavenge oxygen in polymer gel is non-negligible, as is illustrated in figure 6b. With optimized scavenger concentrations (4.65 mM [30]), scavenging time is on the order of 1 hr. Hence, it can be recommended that polymer gels are not irradiated prior to this wait time.

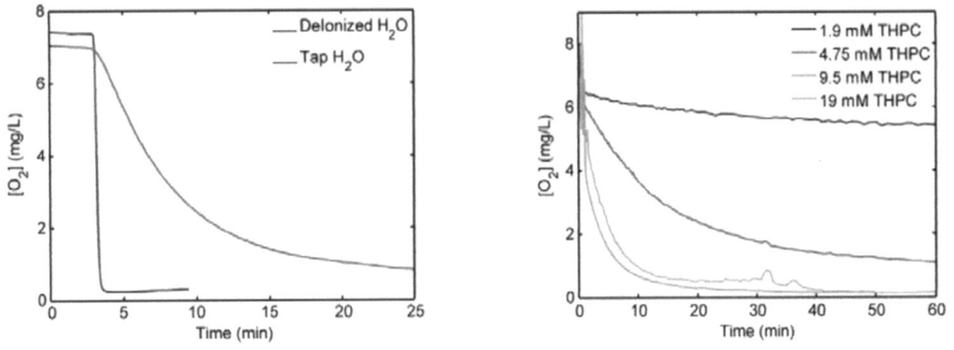

(a) (b)

Figure 6: (a) Rate of oxygen scavenging of THPC for tap an de-ionised water. (b) Rate of oxygen scavenging of varying concentrations of THPC in full gel solutions [30].

6. Effects of THPC Concentration on Polymer Gel Dose Response

The concentration of THPC present within a polymer gel not only affects the oxygen scavenging rate, as seen above, but also the dose-response of the gel. Figure 7 illustrates the effects of THPC concentration on polymer gel dose response. Clearly, too low a THPC concentration will result in not all oxygen being scavenged from the gel and the resultant dose-response will be either flat (unreactive gel) or will exhibit a threshold effect, where gel response is observed only after a certain dose. At a given THPC concentration (4.65 mM) the threshold effect will disappear and maximal polymer gel dose-response will be observed. Above this concentration, however, gel dose-response appears to weaken, likely due to the additional reaction of THPC with gelatin, causing increased gel crosslinking.

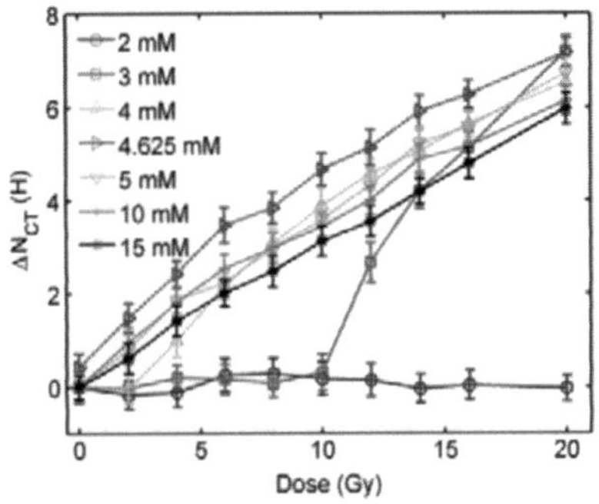

Figure 7: Effect of THPC concentration on nPAG polymer gel dose response [30].

7. Effect of temperature on gelation

Gelatin is composed from small chains of amino acids extracted from natural sources such as the porcine skin. The physics and chemistry of gelation is a field of research in itself and the reader is referred to this literature for further details [31-36]. Here, we only provide the general concepts and their impact on the performance of the dosimeter. Gelatin chains are water soluble such that they dissolve readily in water heated to ~ > 35°C. Upon cooling, the chains form coils and undergo a helix